Evaluating Remote Nuclear Reactor Control

Penney J. Simmons

Abstract

This dissertation provides a comprehensive exploration of the feasibility and potential benefits of remote operations as a control strategy for advanced nuclear reactors. While remote monitoring and control have been implemented in other industries, the operational feasibility and guidelines for implementing remote-operated control rooms for nuclear facilities remain largely unproven and scarce. Therefore, this research offers valuable insights for the industry as it navigates the future of nuclear power plant operations.

The dissertation is organized into six chapters, with each chapter building on the previous one. It begins with a review of state-of-the-art technology for modern control rooms in various industries, including nuclear. The subsequent chapters discuss remote operations in other industries and ways the nuclear industry can implement remote operations. The research then provides a demonstration of remote operations at an advanced nuclear reactor company, highlighting the technical feasibility of remote operations.

Moreover, this dissertation offers two more original contributions. Firstly, it presents a human factors-based study that sheds light on the challenges of remote operations and emphasizes the significance of appropriate training, communication, and workload management to ensure successful remote operations. Secondly, it provides a comprehensive set of criteria that advanced reactor companies need to consider before implementing a remote control room. These criteria encompass safety, cybersecurity, human factors, training, and licensing, serving as a guideline for advanced reactor companies planning to adopt a remote control operational strategy.

Overall, this dissertation provides valuable contributions to the field of nuclear energy. By identifying the key criteria that must be fulfilled for successful implementation of remote operations, this research offers practical insights that can help industry professionals make informed decisions about the future of nuclear power plant operations.

Contents

List of Figures

List of Tables

Nomenclature

ARCO	Advanced Reactor Control and Operations Facility
ATCT	Air Traffic Control Towers
CBP	Computer-based Procedures
CIET	Compact Integral Effects Test
ConOps	Concept of Operations
DMCR	Digital Main Control Room
ETU	The Engineering Test Unit
FAA	Federal Aviation Authority
FHR	Fluoride-salt cooled High-temperature Reactor
HMI	Human-Machine interface
IC	Instrumentation and Controls
ICS	Industrial Control System
IDMZ	Industrial Demilitarized Zone
IoT	Internet of Things
IRB	Institutional Review Board
IT	Informational Technology
LTD	Local Test Director
LTE	Local Test Engineer
LWR	Light Water Reactor
MCR	Main Control Room

NPP	Nuclear Power Plant
NRC	Nuclear Regulatory Commission
OPC UA	Open Platform Communication United Architecture
OSS	Operator Support System
OT	Operational Technology
P&ID	Piping and Instrumentation Diagram
PLC	Programmable Logic Controller
RCSS	Reactor Control and Shutdown System
RO	Reactor Operator
RO	Remote Operations
ROC	Remote Operations Center
RPS	Reactor Protection System
RT	Remote Towers
RTD	Remote Test Director
RTE	Remote Test Engineer
RTM	Remote Tower Module
SCADA	Supervisory Control and Data Acquisition
SMR	Small Modular Reactor
SRO	Senior Reactor Operator
STA	Senior Technical Advisor
TD	Test Director
TE	Test Engineer
UAV	Unmanned Aerial Vehicles

Acknowledgments

Completing this dissertation marks the end of an educational journey that spans over two decades, and I would like to take a moment to express my deepest gratitude to those who have made this final hurdle possible.

Firstly, I would like to express my sincere gratitude to my advisor, Professor Per Peterson, for his guidance, support, and expertise throughout my PhD program. His encouragement and mentorship have been invaluable to my personal and professional development, and I am truly grateful for his contributions to my success. I would also like to give a huge shout out to Professor Max Fratoni and the entire TH Lab for supporting and encouraging me when I needed it most.

I am also grateful to the entire team at Kairos Power and I would specifically like to mention Christopher Poresky and Anthonie Cilliers, whose support and encouragement throughout writing this dissertation have been invaluable. I had the opportunity to work with such a talented and passionate group of individuals, and I learned so much from them.

Additionally, I would like to extend my heartfelt appreciation to the team at Idaho National Laboratory, who provided me with the resources and guidance necessary to conduct my research. Their expertise and dedication were essential to the success of my work, and I will miss our weekly calls, which always began with some random and fun catching up. Thank you so much Ron Boring, Thomas Ulrich, Roger Lew, and Anna Hall.

To my best friend, Camille Waechter, for seventeen years, you have been a constant presence, offering unwavering support and encouragement. Our phone calls, venting sessions, and laughing fits have been a source of joy and relief during the most challenging times. And let's not forget our epic Just Dance sessions, which never failed to lift my spirits. Throughout our friendship, you have accepted me for who I am, and have gently allowed me to grow. Your support and encouragement have been an anchor during the most turbulent times, and I am forever grateful for you. There simply aren't enough words to express how much you mean to me. I am deeply honored to call you my best friend.

I would also like to thank my sister, Stephanie, for her constant support and care throughout this journey. Her regular calls and texts to check in on me, as well as her chef-level home-cooked dinners, have been a source of comfort and joy. I am grateful for her love and encouragement, and for being such an important part of my life.

Finally, I would like to acknowledge and thank my parents for their unwavering support and encouragement. Ever since I was little, I remember being curious, with an insatiable urge to solve puzzles and figure things out on my own. In this pursuit, I owe immense gratitude to my parents for always encouraging me to carve my own path in life. They instilled in me the belief that I could accomplish anything I set my mind to and become anyone I wanted to be. Although I now understand that our core being isn't entirely within our control, I realize that it was through their upbringing that I inherited the qualities of a problem solver, kindness, resilience, and strength. Most importantly, they taught me to care deeply about others and the world around me.

It is this deep sense of care that has shaped my aspirations to be someone who contributes to the betterment of the world. It was the values instilled in me during my upbringing that cultivated my compassionate nature, ultimately driving me to pursue a PhD in nuclear engineering. I firmly believe that nuclear power has the potential to make a positive impact on our planet, and I am honored to be a small part of the effort to uplift the nuclear industry during this critical time.

Chapter 1

Introduction

1.1   Background and Motivation

Over decades, all across the world and across industries, important rooms have been built stuffed full of equipment to provide essential controls to monitor and manage complex facilities. These rooms have been designated "control rooms." The control room is a place where information is gathered and displayed, in some cases automated systems perform control functions based on this information, and operators use the information and the controls to take additional control actions. For example, every airplane has a control room where the pilots sit and control the aircraft. NASA's control rooms are iconic, as captured in photographs of people gathered in a large room sitting at different computers, all wearing headsets, coordinating, and presumably watching a rocket launch. Independent System Operating organizations for large electric power grids have similar control rooms. Power plants, be they natural gas plants, solar plants, wind farms, or nuclear power plants, all have control rooms. Throughout time, these control rooms have been filled with different pieces of equipment and operated using unique protocols reflective of what needed to be accomplished to control the facility, whatever it may be. Additionally, these control rooms represent centralized locations where all controls and operations come together. Like many other segments of society, control rooms have evolved over the past several decades with technological advancements.

Modern technology has completely changed how humans interact with complex physical systems. Much has happened in the past few decades with technology development. Display and visualization technology has advanced substantially; look no further than your iPhone or high-definition television. Large and small screen visualization technologies can provide almost real images; the quality is just that good. Wireless communication networks have expanded, including the creation of fiber optic cables for handling and transmitting data allows for network connection in situations where we may be concerned Wi-Fi will cut out, or environmental conditions are harsh. Software has developed substantially, and it more easily allows for wireless communication, gathering and processing of real-time data, and recording large data streams. In addition, computers can more easily interact with sensors, valves, and pumps [41]. This connects to the rising hot topic of the industrial internet of things (IoT), which is the use of smart sensors to enhance industrial processes [5]. IoT allows any device or "thing" with an embedded sensor to communicate wirelessly with the broader world. Machines, vehicles, buildings, animals, and even people can be part of the internet of things. Examples of IoT applications include the ability to see what is happening in our homes when we are not there, to automatically adjust environmental settings (such as heating and lighting) depending on the weather, and to enable smart appliances such as washing machines to automatically switch on when power costs are at their lowest. IoT paired with software advancements means one can develop an entirely digital control room with all the real-time information and controls needed, displayed on a human-machine interface (HMI).

Most important for this dissertation is the connection between the technological advancements described above and the decentralization of the control room. With real-time monitoring and intranet/internet communication, it is now feasible for the control room to be decentralized or in a different location from the facility itself. The history of the control room as a local place where information is gathered and controls are housed to be used by operators began around the 1920s when mass production emerged. The control room was meant to oversee production from a centralized location as production lines became more complex and widespread. A single control room would lead to higher efficiency, better coordination between groups of workers, and allow for more effective responses in emergencies because all monitoring and coordinated decisions came from one place, not from several locations within a factory that may not have all the information to make decisions [6]. The idea was to consolidate control and operations of decentralized people and assembly lines into one centralized location, the control room. But now, with technological advancements, industries are considering the benefits of decentralizing their control rooms. Furthermore, with the Covid pandemic, there's been an increase in interest in decentralization practices. Covid, in some ways, showed people that they could be very effective remotely, and in some cases, it makes more business sense to be remote.

The concept of decentralization and remote operations, where the control inputs to a facility largely come from outside the facility's bounds, is not novel to many industries. Examples include remote operation of the International Space Station, unmanned 'rovers' on different planets, and drones. However, remote operations have not always been possible or desirable for certain industries like nuclear, oil & gas, and aviation, to name a few. At the end of the day, there is a more considerable cost of failure for a physical plant or an airplane than there is for a rover on Mars. The nuclear industry is conservative when it comes to making updates to nuclear power systems. Digital upgrades can be intimidating to people, and even more unnerving is the idea of nuclear power plants being controlled miles away from the facility. But there is now a case for decentralizing the control room in industries that previously would never have considered such a practice, including the nuclear industry.

1.2 Advanced Nuclear Reactors

In the past decade there has been a resurgence in nuclear power development as a new generation of nuclear power plants have been designed, opening the door for discussion of decentralized control operations. Generation IV nuclear power plants or gen IV or advanced small modular reactors (SMRs) are different names for the new designs of nuclear power plants under development in the U.S. These designs will vary in electricity production size from tens of megawatts of electricity to hundreds of megawatts. In addition, they will vary in the type of reactor coolant used. For example, some designs will use light-water coolant, some gas, and some will utilize molten salt. Additionally, SMRs will have higher operating temperatures to increase efficiency and robust passive safety systems reducing accident risks. Ultimately, the goal is for SMRs to provide more design advantages over their predecessors, such as having smaller physical footprints, lower capital costs and operations and maintenance costs throughout the plant's lifetime, and increased safety. Suppose that SMRs can live up to their expectations. In that case, they offer an opportunity for the nuclear industry to become a prominent player in the energy transition needed to tackle climate change targets and energy supply needs. However, new reactor designs require new control room designs and operational strategies to ensure that these reactors live up to their potential. This section will expand on the characteristics specific to SMRs, which opens the door for new operational possibilities.

Small and Modular

The International Atomic Energy Agency (IAEA) defines SMR reactors as reactors capable of generating up to 300 MWe [30]. A typical light water reactor will generate around 1,000 MWe of electricity. The smallness of these new designs is advantageous for economic and safety reasons. Economically, small reactors are smaller in size and simpler in terms of the components used and the overall design, allowing for lower construction costs. SMRs lose out on some economies of scale that apply to larger reactors, but they are still less expensive, lowering the financial risk of investing in SMR technology. Economies of scale can be made up for by using advanced digitization and automation in plant operations [25]. In terms of safety, smaller reactors require less fuel, but more significantly, they allow for passive safety systems. Passive safety systems refer to features that don't need any electrical feedback or operator intervention in the event of an accident scenario [48]. For example, using automation technology and smart, digital I&C, so the reactor shuts itself down in the case when operating thresholds are reached, and initiates passive cooling using natural forces such as gravity, heat transfer, and buoyancy to cool the reactor down to appropriate levels. The smaller size provides flexibility regarding where the reactor goes physically; this ties to the modular part of a small modular reactor. Modularity means that the subsystems of the reactor are manufactured off-site and then these subsystems are installed and connected on-site. Modularity takes advantage of the simplified reactor components of SMRs and reduces construction costs. Due to their smaller power output size and modularity, multiple SMR units will likely be purchased and installed at a site to provide the desired power output. This kind of multi-modular configuration requires new operational strategies than what's used for existing nuclear plants. We will now have a single power plant with multiple reactor units, all controlled from a single control room.

Reactor Coolants

The advanced reactor designs consider a small range of coolants. Water, helium gas, liquid metal, and molten salts are all options being explored by different groups exploring advanced reactor design. They all present various new design problems the industry will have to investigate. For example, molten salts and liquid metals can be corrosive to the reactor's structural components [53]. In addition, the different working fluids will affect the kinds of I&C that can be integrated into the system, presenting some design challenges.

Higher Operating Temperatures

Due to the changing working fluids from water to fluids with different thermophysical properties, such as molten salts and helium gas, SMRs can operate at much higher temperatures than reactors cooled by water (LWRs), where pressure limits the maximum temperature a reactor can operate at. Higher operating temperatures increase the efficiency of the reactor but also create design challenges and new monitoring challenges. For example, more attention may be needed to what occurs as the coolant cools down too much and the chemical compatibility with components at different operating temperatures. Health monitoring of components and performing diagnostics through the control room will become necessary.

Passive Safety

One goal of the new advanced designs is to create much safer reactors than previous generations. One way that SMRs accomplish this is by making use of passive safety features. As briefly mentioned in the section on small and modular, passive safety refers to safety systems that don't require human intervention or electrical power to function but instead use physics and heat transfer properties to remove heat and maintain temperatures within safe operating limits if thresholds during accidents are reached. For example, the phenomenon of natural circulation is a situation where an elevation difference between the center of heat generation in the core, and the center of heat removal above the core, drive flow. As the temperature of the working fluid rises, the hotter fluid rises. The colder fluid lowers due to density differences, creating a naturally circulating loop as heat is expelled to a heat sink. Another passive safety example would be using magnetically latched shutdown blades that drop into the reactor core when electrical power is removed from the electromagnets. Ultimately, passive safety features change how the operator interacts with the reactor during accident scenarios, a change that needs to be considered in the control room and operational strategies.

Digital Control Rooms

New reactor designs open the door for the industry to consider the benefits of applying technological innovation to the nuclear power plant control room. The visual of a traditional nuclear power plant (NPP) main control room (MCR) that comes to mind for many is a large room with multiple panels consisting of buttons, knobs, switches, alarms, gauges, warning lights, and monitors, which together represent a proper analog system. In this MCR, a crew of reactor operators read the gauges and other indicators and perform manual tasks such as adjusting valves and actuating controls. Applying digital advancement to the MCR provides several advantages over analog systems. For instance, computers are capable of managing vast quantities of data and conveying that information to an operator more efficiently. Furthermore, digitization enables automation of control functions. This allows for more flexible operations to accommodate the increasingly more fluctuating electric grid. This allows for more flexible operations to accommodate the increasingly more fluctuating electric grid.

The transition to digital MCRs (DMCR) is reflected in digital human-machine interfaces (HMIs). With digitization, plant information is communicated via digital HMIs. Boards of panels, buttons, gauges, and knobs are replaced with individual computer-based workstations that can provide significantly more information about plant status. DMCRs contain computerized plant monitors and display systems to provide data to the human operators and control systems that these operators use to control all systems and functions of the plant. They can additionally contain digitized monitoring systems, which monitor plant variables and detect and alert the operators when an anomaly is encountered. Advanced reactor designs embrace all digital systems and intend to use these three systems - plant monitors/displays, plant controls, and plant protection - to ensure plant safety and reliability [42]. The use of digital systems raises several questions: what should these HMIs look like, and what information should be included on display? What should be digitized, and what should not? What should be automated, and to what extent? And how do the operators interact with the control room? Together, the answers to these questions define a new concept of operations for SMR plants. Here, concept of operations (ConOps) refers to a clear strategy that details how the control room is set up and how humans interact with the reactor systems through the control room in various operational scenarios. New reactor designs, considerations, and technology mean a new ConOps is needed for SMRs. This creates space for developments the nuclear industry may not have previously considered with previous nuclear power plants, such as remote control rooms.

1.3 Remote Operations and Scalability

Remote operations (RO) can be defined as a situation where control of your system, in this case, the nuclear reactor, comes from outside the nuclear reactor site boundary. Remote operations is a familiar operational concept. As previously mentioned, industries like space and the military have utilized remote operations to control things like rovers, rockets, and drones. We even see industries that once would not have considered RO are now implementing RO as their new operational strategy—for example, the aviation industry is interested in guiding air traffic control using digital remote control towers, with more details presented in chapter three. Specifically for the nuclear industry, operating a power plant remotely does present some challenges. Basic reactor safety functions need to remain local, meaning the reactor can intrinsically enact safety systems without signals from the control room. But plant control and health monitoring can transition to being remote. To do this, one needs incredibly reliable sensors and equipment, high-speed communication methods, data analytics of large amounts of data, real-time automation techniques, and strong cyber-security [57]. Given the technological advancements we have seen, these challenges can likely be met technically. Still, the implementation must be refined and flexible based on the plant and the operational needs.

Remote operation of nuclear power plants represents a complete shift in operations within this industry, which makes some nervous. Nuclear power is intimidating to many people; it doesn't have the same docile reputation as other renewable technologies like wind and solar. Nuclear power is also substantially more regulated than different energy-producing industries. However, nuclear power is needed on the electric grid if we are to truly move away from carbon-producing energy technology (oil, gas, coal). Furthermore, nuclear power must be able to scale similarly to solar fields and wind farms.

Nuclear power is a zero-carbon-emitting energy source that is technologically capable of fulfilling society's rapidly increasing energy needs. However, it currently contributes only 10% of global electricity production, and some countries have consistently discussed phasing out nuclear [16]. But new developments suggest the industry is ready to take a more active involvement in decarbonization. The development of advanced reactors has led some countries to reassess the role nuclear may play in future energy grids, such as the U.S., home to several advanced nuclear reactor startup companies. The legislation also indicates an increase in support for nuclear power financing, for example, the U.S. Inflation Reduction Act, which provides incentives for the production and use of nuclear power [51]. Given these developments, the nuclear industry has thrown its hat in the game as a front runner for supplying the energy needs once provided by oil and gas. However, construction timelines, safety, and maintaining low operations costs will be critical for the industry to scale up how it needs to.

Renewable technologies have been able to grow faster than nuclear power because the technologies can be built almost anywhere and can be built rapidly at lower costs than other energy generation technologies. For example, solar and wind farms can be built in months, whereas a nuclear power plant takes years. Economic and pragmatic issues hinder the scalability of nuclear power. The technical design of SMRs alleviates some financial and practical barriers. Remote operations of SMRs, while keeping basic safety functions local, can further reduce scalability challenges. The ability to control many geographically dispersed SMR units from a single control room could provide a critical path towards increasing the scalability of nuclear power. Hence, the remote operation of an SMR needs to be an operational configuration that advanced reactor vendors strongly consider implementing. The future of nuclear power plant control systems holds several opportunities for making nuclear power plants smarter, safer, more flexible, and more applicable to society's future energy needs. This dissertation will provide baseline case studies to serve as building blocks for implementing remote operations into the value proposition of SMRs.

Problem Statement and Research Objectives

For operational and business reasons, remote operations may be the control strategy used for advanced nuclear reactors. Remote monitoring of nuclear reactors is a capability being employed by researchers; however, remote operations of an advanced nuclear facility have yet to be proven, and guidelines for implementing a remote-operated control room, along with which basic safety functions remain local, are sparse.

The research presented in this dissertation will illustrate the feasibility of remote operations of an advanced reactor concept, demonstrate a case study for the scalability of nuclear power utilizing remote operations, and establishes a concept of operations and criteria for remote operations of an advanced SMR that can be iteratively evaluated.

Original Contribution

This dissertation offers the following contributions:

- A demonstration of the technical feasibility of remote operations of a physical advanced reactor test loop

- A human factors-based study of a remote control room baselining human performance when controlling more than one plant from a single control room

- A list of perceptions from various stakeholders in an advanced reactor company that need to be addressed for remote operations to be an operational configuration

- A comprehensive list of criteria that an advanced reactor company should address in order to implement a remote control room.

Dissertation Organization

This dissertation takes the reader through the modernization of control rooms across industries to a remote operations proof of concept and the implementation of this operational strategy.

- **Chapter 2** reviews the state-of-art technology for modern control rooms for nuclear and other industries.

- **Chapter 3** discusses examples of remote operations in other industries and ways the nuclear industry can implement remote operations.

- **Chapter 4** provides a demonstration of remote operations at an advanced nuclear reactor company, specifically designing a fluoride-salt cooled high- temperature reactor (FHR) SMR.

- **Chapter 5** walks the reader through a human-factors, remote operations ConOps use case conducted utilizing an advanced reactor simulator.

- **Chapter 6** lists perceptions of remote operations from various stakeholders, introduces the remote operations criteria a reactor company should fulfill based on lessons learned, and outlines a path forward for a remote control operational strategy of an advanced reactor.

Chapter 2

Modernizing Control Rooms through Technological Advancements

The control room has been part of an evolving landscape where analog systems are replaced with their digital counterparts. The role of the control room has always been to serve as a centralized place where humans or operators interface with the physical systems they are responsible for. Before technological advancements, the analog control room consisted of physical buttons, switches, and alarm lights. In this control room, operators need to read the gauges and other physical indicators and perform manual tasks such as adjusting valves and pressing buttons using the hard-wired controls. In addition, the operator must have extensive knowledge of the system to make appropriate decisions about changing values [42]. When people refer to the modernization of the control room, they are typically referring to the digitization of the control room from the sensors to the visualization techniques, enabling digital control rooms to organize, manipulate, and display data in ways that are entirely different from their analog counterparts. The control room remains the centralized place where humans interface with a physical system, but the ways they interface with the physical components have changed as we have moved away from analog control to digital control.

This chapter provides background on the modernization of control rooms across industries. Without the modernization and digitization of control rooms and the technological advancements that allowed such modernization, this dissertation topic wouldn't be relevant. Therefore, this chapter begins with a historical review of the technological developments that allowed digitization. Then it will cover standard features of digital control rooms consistent across industries. Finally, digital transformation and future remote control room possibilities will be discussed.

2.1 A brief history of Industrial Control Systems

A basic control system has two pieces. One piece is controlled, and the second piece provides control. The piece being controlled has certain output variables which need to be controlled using input variables that can be adjusted. In its simplest form, a digital control system must observe and change the output variables via new input variables [17]. The digital systems also collect additional data to monitor the system, such as plant health data. An industrial control system (ICS) describes integrating hardware and software with network connectivity to create a digital control room. A typical ICS includes programmable logic controllers (PLCs), supervisory control and data acquisition (SCADA) software, control servers, sensors, and some networking protocol. Industrial control systems are essentially the brain of the digital control room; thus, a description of the components that make up an ICS must be reviewed.

PLCs

Before the 1960s, control systems utilized relays. A relay has two components, a relay coil, and a relay contact. Relays work by using the relay coil that can be energized or de-energized to create a magnetic force that will change the state of the relay contact, which then pulls a switch to an on or off position (Figure 2.1).

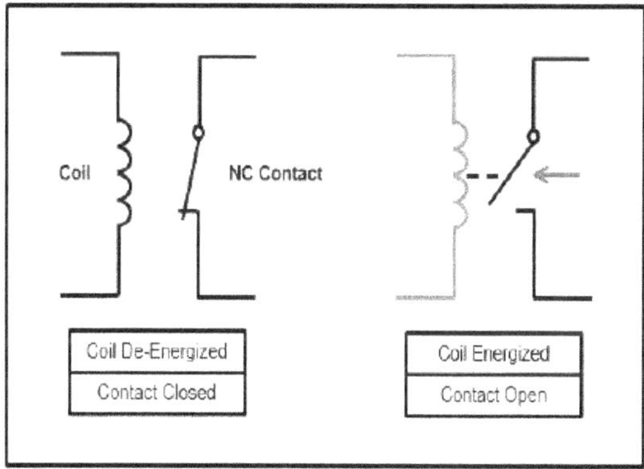

Figure 2.1: Relay Operation with a Relay Coil and Relay Contact [36]

For example, to control a motor, a relay would be connected between a power source and a motor, and an operator would control the motor by controlling the relay to the power source. Each relay contains a minimum of four wires to function properly. But each relay can only control a single circuit; therefore, any equipment needing control would need many relays, creating a logistical nightmare and safety hazards.

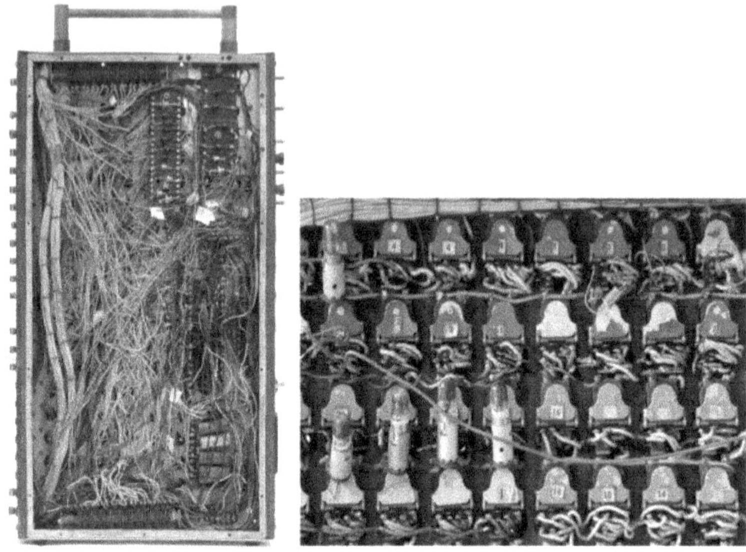

Figure 2.2: Relay Logic Machine (1950) [15]

After relays, control systems used vacuum tubes. A vacuum tube is a device that uses the flow of electrons in a vacuum to create an electrical signal. A metal electrode, also known as a cathode or filament, is heated, causing electrons to emit from the surface. The negatively charged electrons are attracted to the positively charged anode and travel across the vacuum to hit the anode, creating a current. A voltage can be applied to this system, which will control the flow of the electrons, making an on/off switch [32]. Vacuum tube technology was utilized by notable organizations such as IBM to power the first computers, radios, and early long-distance telephone networks. Vacuum tubes can only produce as much power as the current they can create (Figure 2.3). Several vacuum tubes are needed to support large output power, and each filament requires constant and considerable power. Additionally, the filaments degraded quickly. Continuous monitoring and replacement were needed to keep supplying the current required for control systems. In 1968 the invention of special control computers known as programmable logic controllers (PLCs) were developed and changed the way control systems work [24].

A PLC is an industrial computer without a mouse, keyboard, or monitor (Figure 2.4). Fundamentally a PLC takes in inputs, executes instructions given in a programming language rather than a hardwired electrical signal, and then puts out outputs. For example, a PLC monitors the status of switches and sensors and then communicates the status, and the operator can send signals back to the PLC and the hardware through the program logic.

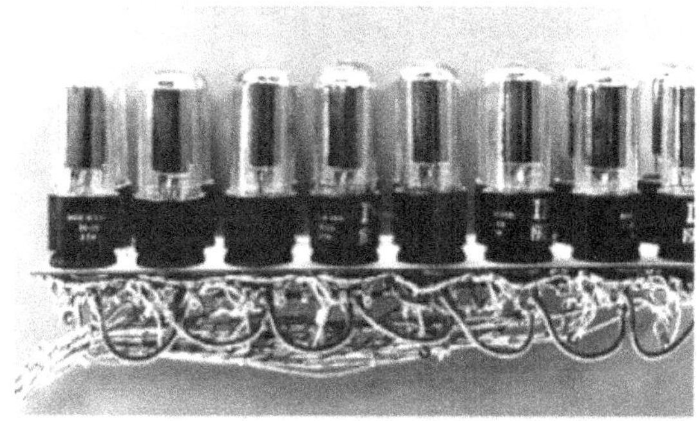

Figure 2.3: IBM Vacuum Tube System, 1940s [26]

With one piece of technology, massive numbers of electrical wires, relays, and vacuum tubes were replaced with a box with enormous computing power. The automotive industry led by Henry Ford began one of the first industrial control systems we think of today. GM utilized PLCs to revolutionize its factory assembly line and create the first instance of a digital control system. The PLC drove further technological developments because it was invented before the personal computer and sophisticated networking. The PLC set in motion a cascade of events that led to today's industrial control systems.

Figure 2.4: Toshiba PLC [59]

2.2 Supervisory Control and Data Acquisition (SCADA)

SCADA is a software tool that integrates with a PLC to create a complete digital control room experience. A PLC, on one end, communicates with an array of devices such as sensors, pumps, and machines; on the other end, it communicates with a computer with SCADA software. SCADA uses a server-client architecture. The SCADA software takes in the data from the PLC, processes it, and stores it in a server. A SCADA server is responsible for data acquisition and management. The client on a computer connects to the server to gather the data and display it to the human operator [27]. The three main pieces of SCADA is to supervise the data in a graphical representation, control processes, and acquire real-time data.

Networking and Communication Protocols

For the industrial control system to work, the equipment must communicate with the PLC, which then needs to be able to communicate to the computer with the SCADA software. This communication is referred to as a network communication protocol. Networking refers to the connection between two or more devices where information is exchanged. Each device is known as a node; how they communicate is known as a link. A node could be a computer, a server, a smartphone, a gaming console, etc. Links require some medium like wires, optical fibers, or air that data can be sent through between nodes. Figure 2.5 shows the most straightforward network of nodes and links [64]. The nodes are circles A, B, and C. The two lines between them represent the links. A can transfer information to node B, and node C can transmit information to node B, and node B can share information with nodes A and B. B must act as an intermediary to send data from node A to node C or vice versa. B must be a cooperating node between A and C because a direct link doesn't exist between them.

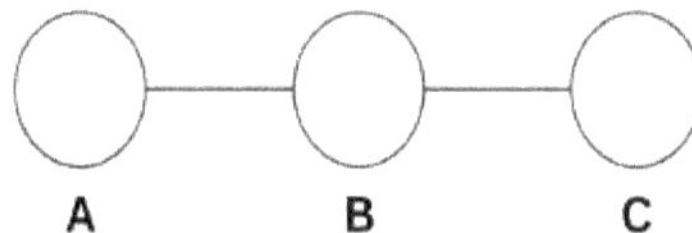

Figure 2.5: Simple Node-Link Network

For networking protocols, a common way of linking nodes without a physical link is to use an Internet Protocol or an IP address. An IP address is expressed as a set of twelve digits separated into four segments known as bytes. For example, "xxx.xxx.xxx.xxx," where each "x" is a number. These twelve digits make up an IP address. The left part of the address is the network-identifying component (the network ID), and the right part is a device-identifying component (the host ID). The network uses the network ID to communicate data to the target device using the host ID. For example, if you want to use the internet on your computer, your computer first connects to a local network that is connected to the internet. Then the internet sends and receives information to your device using the host ID. IP addresses allow devices to find each other anywhere in the world [63]. Another possibility to connect nodes would be to utilize an ethernet protocol. This simple node-link network visually displays the ability to transfer data; however, transfer speed, accuracy, and reliability of the transfer networks can be just as critical in many cases. Additionally, node systems are usually more complex than described in the Node-Link network example. Different nodes with different requirements for links need to be connected. For example, node A communicates wirelessly, but node C requires a wired connection. Additionally, node A is connected to child nodes A1, A2,... An that can only communicate with each other and with node B, but it is desired that the node C device receives all the information from node A [64]. We now have a more complicated network infrastructure that is more representative of an IoT set of devices that are all connected and must work together.

While communication protocols may seem like abstract software technology, almost everyone is familiar with some protocols, such as HTTP and ethernet. HTTP refers to Hyper Text Transfer Protocol, a protocol of the internet. If you've used the internet recently, you may have browsed a website with an address that begins http:// or https://. Essentially an individual asks for a specific webpage to load, and the computer system responds by fulfilling that request and displaying the website. The difference between HTTP and HTTPS is that the latter uses encryption to protect any information sent between a client and server. Websites can be given a secure sockets layer (SSL) certificate to encrypt information sent between your computer and a server. If a website has an SSL, it can utilize a HTTPS, indicating to users that their website will protect information between the website and the user [34]. Ethernet is another communication protocol many have encountered. The most common ethernet protocol is TCP, Transmission Control Protocol, which shares data between devices all connected to routers and switches. Two other common examples of connecting devices are a local area network (LAN) and a wide area network (WAN). A LAN is a private and localized network connection. A WAN provides communication over more considerable distances from the distance of a different building, a city, a region, or even greater distances. WANs can be used to connect multiple LANs. For example, Wi-Fi is the most well-known wireless LAN, and cellular networking such as 4G or 5G is a familiar wireless WAN [64]. There are other types of communication protocols that all communicate data slightly differently, at a different scale, and with varying layers of encryption and security. One will select the protocol that optimizes their specific industrial control system. But an IoT system will likely have a combination of networking protocols to allow multiple devices to communicate with one other and over different distances.

OPC UA Industrial Standard Protocol

Along with the networking protocols described above, industrial industries are using a protocol known as OPC UA for the secure and reliable exchange of information. OPC UA stands for Open Platform Communication United Architecture, a standard industrial communication protocol that SCADA systems can use to enable different devices operating with other protocols to communicate together. It is open-source, secure, and scalable software, making it valuable for industries. Generally, real-time data comes from your machine, it is then collected in a PLC, and then OPC UA connects to that PLC, and for every data point, a node is created in OPC UA. Then OPC UA sends the nodes to a SCADA computer device, allowing one to see the nodes containing the data. For example, in the Basic SCADA diagram below, the red lines represent the OPC UA communication between the PLC, server, and clients. The black lines represent any communication protocol the devices and sensor use to communicate data to the PLC (typically one of the protocols described in the previous section). Many industrial systems have multiple sensors and equipment that send data to a PLC. OPC UA doesn't care what communication protocol the hardware uses to reach the PLC [39]. Once connected to the PLC, OPC UA operates as a language translator and brings the separate information together through the server and to the workstations.

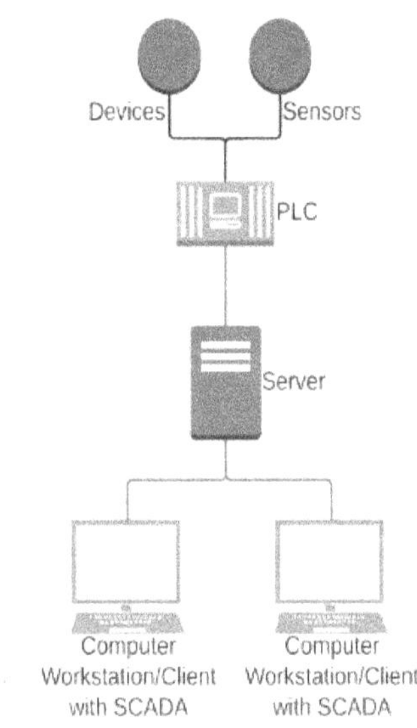

Figure 2.6: OPC UA Communication Protocol

2.3   Features of the Control Room

With the developments of communication between software and hardware technologies, the development of IoT systems, and the growing complexity of industrial processes, control rooms have been updated from their analog counterparts. The intentional use of networking techniques and advanced displays will only increase as society advances software and hardware technology. While every control room is different, digital control rooms maintain some consistency. This section will explore the common features of present-day digital control rooms.

HMI Development

Human-machine interfaces (HMIs) play a crucial role in supervisory control and data acquisition (SCADA) systems by providing users with a visual representation of data collected from programmable logic controllers (PLCs). Thanks to digitalization, plant information can now be communicated through digital HMIs, allowing operators to interact with computer-based workstations instead of traditional panels with buttons, gauges, and knobs. With new and flexible designs, HMIs enable operators to control and monitor industrial processes, track trends, visually display data, and receive alerts on their screens for potential errors [42]. Advancements in display and visualization technology have also greatly improved the user experience, as evidenced by high-definition televisions and smartphones. Modern visualization displays can create almost realistic images, making it easier for operators to process and understand the data presented to them. For example, Figure 2.7 shows a side-by-side comparison of an analog NPP control room and an updated control room test facility equipped with modern visualization displays.

By centralizing and digitizing data, digital control rooms optimize the interaction between humans and systems. Operators no longer need to continuously walk around various panels or manually review data streams. Instead, HMIs visually display pertinent data and controls in a centralized space, allowing operators to efficiently manage processes in a more intuitive way. Effective HMI design is therefore critical, as it directly impacts the operator's ability to fulfill the objectives of the control room.

Figure 2.7: Kola NPP and the Human System Simulation Laboratory [35] [10]

Across industries and disciplines, when a control room is needed, researchers and designers work towards developing the ideal digital control room and HMI system to better enable the control room to fulfill its purpose. However, simply because a system is digital does not automatically make it better, more practical, or safer than the analog counterpart. Various technologies lay the groundwork for digital control rooms to increase the value proposition of whatever they control, but work must be done to see that through. This work comes from the actual design of the HMI.

Designing an HMI requires careful consideration of layout, navigation, colors, data shown, how the information is displayed, and an overall vision of how the human interacts with the screen. There is no one size fits all approach to HMI and control room design. The designs will be different for every system because systems have different requirements. For example, the interface that drivers interact with in a Tesla vehicle differs from an interface one might use to control a power plant. The interfaces will look different, contain different information, and the human may interact with the interface differently.

Operator Support Systems

Operator support systems (OSS) refer to tools built into the HMIs that collect, optimize, and display information. While the design of interfaces varies, we see some similarities across interfaces through OSSs. The rest of this section describes some of the more common operator support systems across industries in new digital control rooms. The actual design of the OSSs will differ because there is no one size fits all approach to the design of a control room. Still, their basic structure and use are essential to minimize adverse effects and maximize benefits.

OSS - Soft Controls

In the digital control room, human-machine interfaces (HMIs) incorporate soft controls, which are input-based interfaces that use software to connect physical objects to display systems [65]. Soft controls enable operators to interact with the system by clicking buttons on computer-based screens, which change the state of physical objects, such as turning a pump on and off. In contrast, in an analog control room, operators would have to physically switch a hand switch up or down to change the hard-wired signal to the pump. While a single switch in an analog control room would typically control one specific object in a specific manner, in the digital control room, a single button can produce different outcomes based on how the soft control is programmed.

An example of a soft control representation is shown in Figure 2.8. This image comes from the HMI for Advanced Reactor Control and Operations Facility (ARCO), a digital control room prototype developed by researchers at UC Berkeley. ARCO uses digital HMIs to control the compact integral effects test (CIET), a scaled-down test facility used by researchers studying the SMR concept of the FHR. In one of the ARCO HMIs, there is a soft control that users control a pump on CIET. The left image shows the pump as off with a 0Hz value. The image on the right shows the pump is on, indicated by the blue highlight of the "activate pump" button, and the pump visually reads and shows a value of 42Hz. The user typed and entered that value in the white box to the right of the pump figure, and the soft control connection led to the control of the pump.

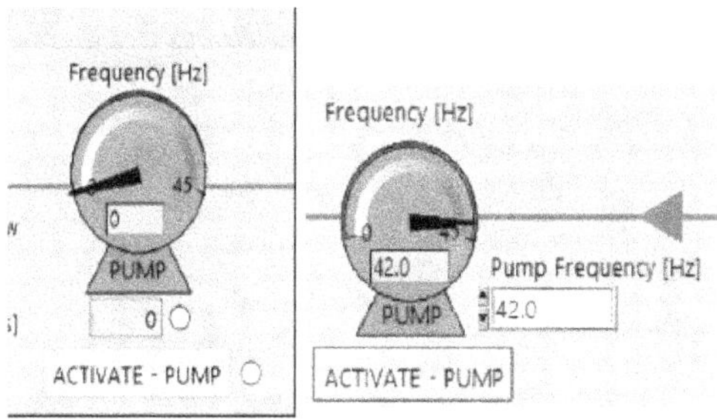

Figure 2.8: ARCO Pump Icons Showing Soft Control Functionality

OSS - Advanced Alarm Systems

With the advent of digitized control systems, computer-based alarms have been integrated into HMIs, providing a significant improvement over their analog counterparts [66]. Alarms have always been a critical component of any control room, serving as the primary means of alerting operators to attend to an abnormality. In analog control rooms, alarms typically took the form of a siren, which would produce an audible and visual alert to indicate an error, leaving the operator to diagnose the problem and address it.

With the integration of alarms into digital control rooms, operators are now provided with advanced alarm systems that actively assist in determining the cause of the problem. These systems are integrated into the HMI, providing a visual display of errors that require attention. The advanced alarm systems allow for the use of colors to indicate different levels of importance. Red, for example, is used to signify an urgent situation that requires immediate action, while yellow suggests that attention is needed, but the situation is not yet critical. In the past, green has often indicated that an automated protection action has been taken to prevent a severe condition from occurring.

In Figure 2.9, we see an HMI that represents various components of an actual physical system. One of the indicator bars has a yellow box around it, with red and orange bars indicating that the temperature level of TC-301 requires operator attention. The use of colors in this case indicates that the situation is not immediately critical, as the temperature value is in the orange zone and not the red zone. This highlights the advanced capabilities of digitized alarm systems and their ability to provide more intuitive and user-friendly information to operators.

Another example of an alarm is depicted in Figure 2.10. This image comes from the ARCO HMI. We can see that the primary flow value is highlighted in red. This indicates to the operator this value has reached a threshold and needs to be addressed by the operator immediately.

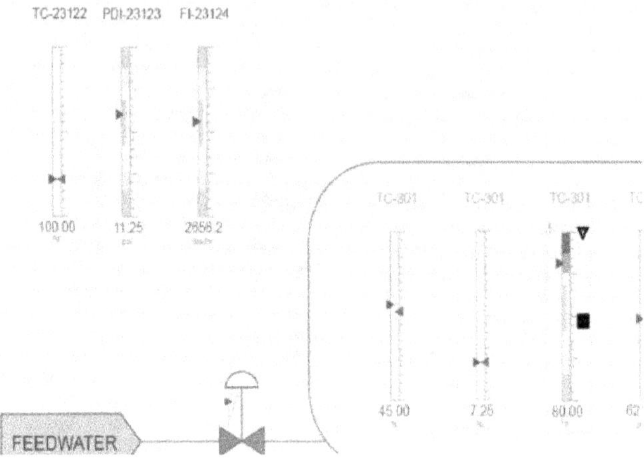

Figure 2.9: Rockwell Automation Style Guide Alarms [**Automation˙undated-qp**]

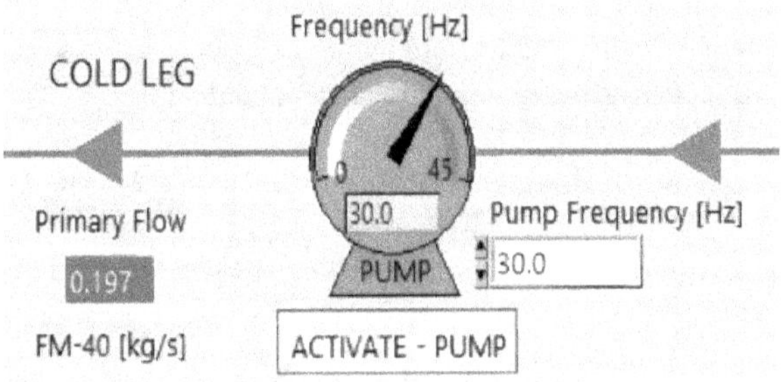

Figure 2.10: ARCO Alarm State Example

In digitized control systems, every aspect of the control room is redesigned. The information displayed to operators can now be presented in ways that not only alert operators to a problem but genuinely aid them in diagnosing the cause of the problem and prioritizing the order for addressing various alarms, both of which benefit the smooth operations of the control room, limit human error, and increase safety and efficiency.

OSS - Computerized procedures

In industries that control complex systems such as power plants, strict procedures dictate all actions. For years, paper-based procedures (PBPs) have been the norm, requiring operators to navigate through the manual procedures step-by-step. However, with technological advancements, computer-based procedures (CBPs) have emerged as a viable alternative

[52]. These digital procedures can take various forms, from direct digital copies of their paper predecessors to more interactive formats that guide operators through each step. One of the primary benefits of CBPs is the reduction of human error, as they can track operator actions and alert them if they miss a step, ensuring the procedure is correctly followed. CBPs can even collect plant data in real-time and inform the operators, improving efficiency and safety [42]. The design of CBPs, much like the rest of the OSS, is dependent on the HMI design and the specific requirements of the control room to ensure the highest level of safety and efficiency.

2.4 Digitization, Digitalization, and Digital Transformation

In previous sections, we've discussed the advantages of digitization, which is the process of replacing analog control with digital technology [9]. Digital control room systems have numerous benefits that can significantly enhance the efficiency and quality of the control room. For instance, digital systems can handle a vast amount of data and communicate it in real-time, integrate data and control logic, incorporate digital visual technology with aesthetically pleasing displays, and provide cost-effective maintenance through readily scalable and replaceable digital components [62].

Digitization changes the way users interact with the system and with each other. The change in the interaction between the user and system that comes from digitization is known as digitalization. Digitization provides a new technological infrastructure for industries to improve their operations. Digitalization is the actual improvement or adjustment of processes using digital technologies. Both digitization and digitalization are the backbones of digital transformation, which as the name suggests, is the fundamental transformation of the day-to-day industries leveraging digitization. Digitization, on its own, offers benefits, but a total digital transformation provides even more compelling benefits.

Digital transformation is often defined as using digital technologies to completely re-think and create new ways of performing day-to-day processes [9]. Digital transformation allows for new business and operational strategies by utilizing technological advancements that weren't possible several decades ago. Digital transformation can mean many different things depending on the industry, but the commonality is a fundamental change in how things are done. This transformation has given rise to various possibilities in the control room, including predictive health monitoring for improved maintenance, autonomous operations, separation of protection functions from control functions, and remote operations.

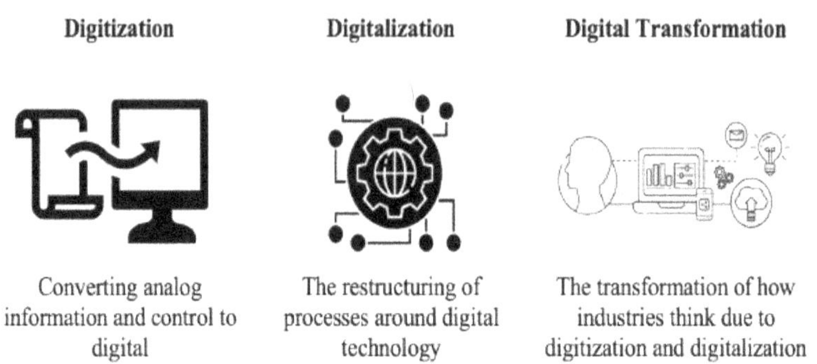

Figure 2.11: Digitization, Digitalization, Digital Transformation

Predictive Health Monitoring and Maintenance

The digital transformation of systems occurs when traditional thinking is scrapped, and new ideas for transforming systems are embraced, ideas that harmoniously utilize digital technologies and operation strategies to make systems highly efficient. Operations and maintenance practices tend to be the most significant financial drain on a system across industries. Digitization and digitalization can minimize maintenance and operations costs if utilized strategically. The digital control framework is such that technology allows for communication between hardware and software located on devices. Sensors are now capable of obtaining information about the condition of the equipment. That information from the hardware is passed through software that can analyze the state of equipment and predict when maintenance will likely be needed. Operators can then plan for this maintenance period [64]. Digital transformation can optimize maintenance practices leading to better business results for facilities.

The Journey Towards Autonomous Operations

Automated and autonomous are not synonymous, though they are often mistakenly considered the same. Automated refers to action taken by technology based on a set of predefined rules programmed by someone. Automated systems make routine or repetitive processes easier. For example, a garage door opening when approaching your home is an automated system. Automated systems can replace tasks once performed by humans [12]. Autonomous systems make their own decisions based on training and learning that allows them to adapt, identify problems, and act independently. An autonomous system can respond and adjust behavior. An example of an autonomous system is a self-driving car. The car can sense its environment, determine what actions to take, such as switching lanes or breaking, and avoid harmful situations. Semi-autonomous systems are a middle ground between automated and fully autonomous. Back to the car example, semi-autonomous refers to features like lane keep assist and a forward collision braking system [8]. The car can make some choices based on what it perceives from its environment, but not all choices. Digitization and digitalization made much more complex automated systems possible. Digital transformation ideologies sparked the idea of autonomous systems. For future control operations, it is conceivable that most control actions will become automated, and control systems will become self-reliant and can adjust and make decisions based on feedback.

Protection and Safety Functions

With technological advancements, digital transformation, and the ability to create autonomous systems, it is now possible to design complex systems that separate protection and safety functions from control systems. This means that while the protection and safety systems remain local, the control functions can be managed remotely. For instance, in traditional interactive human control systems, such as those found in automobiles, power plants, and rockets, operators need to intervene to ensure safety. This is called a human-in-the-loop control system [38]. Since protection systems typically require human intervention, they are often integrated with control systems, leading to the risk of human error. The good news is that digital and automation tools offer an alternative to human-in-the-loop protection systems. By separating protection systems from control systems, the protection system can be integrated locally within the system, while controls can either remain local or be decoupled from the system. We've seen this approach being used in self-driving vehicles and SpaceX rockets. For instance, self-driving cars integrate high levels of autonomy, enabling them to perform safety functions, such as automatically breaking when sensors and automation signal a potential collision. Similarly, the SpaceX Falcon 9 rocket has an automated safety system that uses GPS and sensors to track the rocket's position and signal an automatic self-detonation if violations are found [45]. This separation of protection systems from control systems is especially relevant for advanced nuclear systems, where keeping the protection system within the reactor reduces the risk of human error and the need for operators and control systems to remain on-site.

Benefits and Concerns

Any operational change will have benefits and issues that must be addressed. The movement towards digital systems, implementing operator support systems such as computer-based procedures, using predictive maintenance and automation, and implementing remote operations capabilities all come with benefits and concerns. Many benefits and issues overlap because their common denominators revolve around digitization and digitalization. For example, digital control systems lead to reduced staffing levels, the use of human-machine interfaces, the addition of operator support systems, some level of automation, and new communication protocols between operators within a control room and between the control room and the system. Furthermore, digital and automated systems can separate safety functions from control systems, providing further safety benefits. Table 2.1 outlines the potential benefits and concerns of each system change due to digitization.

2.5 The Case for Remote Operations: The New Control Room Possibility for Nuclear

The advancement of technology has eliminated the physical limitation of having control rooms on-site of what the room controls. Advanced displays, live data streams, data analytics, and networking protocols have made it possible for remote operations to control systems or tools beyond their typical operating boundaries. While remote operations have been utilized in certain industries such as the space industry and military for decades, many sectors have been slow to embrace this technology. However, the global pandemic caused by Covid-19 has forced industries to take remote options seriously. While remote work is different from remotely operating a power plant, different factors are involved with differing consequences, digital transformation is about moving away from traditional tropes and finding new solutions to a fluctuating workforce and increased demand for operational excellence. Digital transformation made remote control possible, and it is time for industries to take it seriously as an operational strategy. As such, this subsection will describe two remote technology options and make a case for remote operations as a value proposition addition for future nuclear power plants.

Remote Monitoring and Remote Control

Those responsible for industrial facilities or equipment have a vested interest in optimizing gains, whether in the form of increased efficiency, safety, or economic returns. Advances in digitization have made it possible to collect extensive data from a range of devices, including sensors and cameras. This data can be used to inform decision-making, leading to improved outcomes. Additionally, the development of mechanisms for transmitting data from hardware to software has been instrumental in enabling remote operations.

It is worth noting that remote operations can take different forms. Remote monitoring, for instance, involves an operator monitoring the system from a location outside the facility's boundaries. In the event of any issues or necessary adjustments, on-site personnel must be contacted to take control actions. Facilities that rely solely on remote monitoring tend to maintain human-in-the-loop protection and safety systems. For example, in 1999, Mitsubishi Power built a remote monitoring center in Takasago, Japan. In 2001, they built a remote monitoring center in Orlando, Florida, and in 2016 a third center was opened in the Philippines. A fourth facility opened in Nagasaki in 2019. All four facilities remotely monitor over 30 GW thermal power generators and over 150 turbines [43]. These centers monitor the overall status of all the plants and turbine systems by using predictive analytics software and real-time performance calculations to optimize O&M costs [40]. The remote monitoring centers enable more optimized operations within the local control room; however, all control of these facilities remains within the local control rooms.

It is important to distinguish remote monitoring from remote control, as the latter represents a more advanced and comprehensive operational concept. In a remote control room, controls for a system come from outside a defined boundary, and the local control room is not the first line of operations. Remote control rooms offer a range of benefits, including increased efficiency, safety, and scalability, all of which can positively impact a company's bottom line. Furthermore, a decoupling remote control strategy from protection systems can enable even greater gains in these areas. While there are numerous examples of remote monitoring centers, remote control rooms offer additional benefits and are worth exploring in more detail. Chapter 3 will provide specific examples of remote control rooms, demonstrating the many ways in which remote operations can improve operations and contribute to a company's financial success.

Economic Benefits of Remote Monitoring and Control

Specific cost reduction numbers resulting from the implementation of remote control rooms are challenging to find across all industries, this could be due to the proprietary nature of such information. Additionally, enterprises may still be unsure of the exact monetary benefits of remote solutions. Nevertheless, there are indications that there are clear financial advantages. For example, General Electric's Remote Monitoring and Diagnostics services claim that their platform can reduce power plant maintenance costs by up to 30% through performance analytics measures [21]. Furthermore, digital transformations often lead to optimized equipment usage, diagnostics, and staffing reductions, resulting in lower operational costs. Despite the limited documented cost benefits, many companies are still implementing remote monitoring and control solutions.

As remote monitoring and operations become more prevalent in industries, companies are working to understand the total economic gains from future remote implementations. Allied Market Research conducted an extensive study in 2021, which estimated the Remote Monitoring and Control Market in the U.S. at $7.9 billion, with a global market estimated at $25.8 billion, expected to grow to $43.6 billion by 2031 [4]. The market is segmented by type, with monitoring and control options available. Allied Market Research predicts that remote control will experience more significant growth by 2031, doubling compared to monitoring. Increasing automation and the growing need for industrial mobility, scalability, and operational excellence are the primary drivers of this growth. As a result, there is a lucrative opportunity to participate in the remote market, particularly in the power sector.

2.6 Summary

The modernization of control rooms in various industries, including the nuclear industry, has been made possible through the use of software-based control systems. Programmable Logic Controllers (PLCs) and Supervisory Control and Data Acquisition (SCADA) systems have replaced older, manual control systems. These software-based systems allow for more efficient and accurate control of various plant operations, leading to increased productivity and improved safety. Additionally, networking protocols, such as Ethernet and OPC UA, have made it possible for these control systems to communicate with each other and with other devices within the plant, allowing for better data management and control. These technological advances have led to the consideration of remote operations for control rooms in the nuclear industry and other industries.

The nuclear industry has an advantage over other sectors in the power industry, such as oil and gas, thanks to Small Modular Reactors (SMRs). These new reactor designs allow for the consideration of optimal remote operations implementation strategies from the outset of the development process. Retrofitting existing plants for digitization can be a challenge, as remote operations are not easily integrated into a retrofitting approach. Given the high stakes involved in the nuclear industry, any changes in operations at any level of the plant must undergo an iterative design process to ensure plant safety. This includes changing the types of sensors used, evolving control room strategies, and evaluating protection systems in conjunction with remote control systems. The nuclear industry cannot afford another accident, big or small, due to its precarious position in the energy sector and public perception. As industries, including the nuclear industry, face the million-dollar question of whether to manage their control room on-site or offsite, the answer will increasingly be offsite. To tap into the future market, the control room should not only be for remote monitoring but for remote operations as well.

Table 2.1: The Benefits and Issues Associated with Digital Control Systems

System Change	Benefits	Issues
Digital Control Systems	• Reduces staffing • Real-time data acquisition • Fault Monitoring • Separate protection from control systems	• Cyber attack vulnerability
Human Machine Interfaces	• Present plant info intuitively • Improve task performance • Help locate and fix faults	• Cyber attack vulnerability • Challenges in displaying useful information
Operator Support Systems	• Can be tailored to the user's needs • Help the operators identify problems • Reduce human error • Aid operators in performing complex tasks	• Cyber attack vulnerability
Application of Automation	• Opportunities for task support • Decrease operator mental stress and demand • Aid in fault detection and mitigation • Separate protection from control systems	• Cyber attack vulnerability • Potential decrease in situational awareness
Separation of Protection Systems from Control Systems	• Reducing human error • Decoupling protection and control systems	• Cyber attack vulnerability • Potential decreased situational awareness
Communication Protocol between Operators	• More defined roles for operators	• Less direct operator communication
Control Room to System Communication Protocol	• More system data provided to operators • Optimize human-machine labor division	• Cyber attack vulnerability • Potential network connection loss

Chapter 3

Remote Operations in Industries: Learning from Aviation, Drones, and Mining

Despite historical advances in digitization and automation, certain industries, such as the nuclear industry, continue to opt for on-site control of power systems for various reasons. While safety and apprehension are two reasons for maintaining on-site operations practices, the current fleet of nuclear reactors doesn't lend itself to a remote operations strategy. The existing nuclear power plants were constructed with on-site control rooms; in all cases, the control rooms were originally fully analog. These control rooms were situated very close to the reactors to reduce the lengths required for cable runs and required specialized ventilation systems to protect operators if radiative material is released during an accident. After a fire occurred in the cable spreading room of the Browns Ferry plant in 1975, plants were required to add a redundant remote shutdown panel where operators could shut down and monitor the reactor in the event the primary control room became inoperable. The remote shutdown panel is the only instance of any remote control for the existing fleet.

Efforts have been made to retrofit the control room of these existing plants to be more digitized. Retrofitting allows for upgrading the plant's instrumentation and control systems, as well as the control room itself, by implementing Human-Machine Interfaces (HMIs) that feature soft controls and, in some cases, additional operator support tools [62]. However, there is no practical approach to retrofit a plant to include a remote-control room. Furthermore, as the energy grid becomes increasingly variable and unstable, power providers are now utilizing a digital suite that includes advanced sensors, data analytics, and visualization technologies to transform their control rooms into spaces that provide optimal operations and maintenance for the specific plant and the electric grid. Digital technologies allow for power plants to not only be monitored remotely but controlled remotely as well. As a result, the nuclear industry can take advantage of the digital suite and reap the economic and operational benefits from remote control of these new reactors. Furthermore, the move to digital allows for a decoupling of nuclear reactor safety systems from control systems.

Following NRC regulations, reactor protection systems are designed to initiate control rod insertion to rapidly shutdown the reactor if unsafe limits are approached, with the ultimate goal of preventing the release of radioactivity [22]. Every nuclear plant has operating limits, and the RPS is designed around these limits. Similar to the control system of the previous generation of nuclear reactors, the reactor protection systems are also analog. Predetermined actuation setpoints work with analog circuits (typically based on relays) that take in input signals from process sensors to create some reactor trip logic. Typically, the trip logic removes power from the control rod drive mechanisms and the rods drop into the core [67]. The reliability of the RPS is critical. With enhanced digitization, the communication between hardware and software has improved reliability and additional safety features can be incorporated into the reactor protection system. Meaning, that human intervention is no longer needed to anticipate and mitigate off-normal conditions, nor for actuating safety systems. Thus, in a remote operations scenario for nuclear power plants, the protection system can be separated entirely from the control system, further increasing the case for remote operations. Should anything go wrong with the communication between the remote control room and the reactor, the reactor will have an automatic RPS and intrinsic technical aspects for keeping the reactor within safe operating limits without human intervention. The goal for future power plants appears to be autonomous plants remotely monitored and controlled from remote operations centers that can adjust to the energy grid's needs. The transition to autonomous remotely operated NPPs with decoupled autonomous safety systems would be revolutionary for the nuclear industry.

Advanced nuclear power plants offer a promising opportunity for the industry to embrace technological advancements and enable accelerated decarbonization of energy supply. To support these advanced reactor designs, the industry is exploring digitization and considering the idea of autonomous and remote control systems. However, achieving these operational goals will require careful consideration and planning during the design process, as retrofitting or adding on remote operations will not be feasible. The industry needs to prove the concept of operating a nuclear power plant off-site rigorously, as it is a highly reserved industry that is often viewed skeptically by the public. Therefore, defining a set of criteria for remote operations to be considered is essential, and the industry can draw insights from other industries such as aviation, military, and mining. By studying their use cases for remote operations, the industry can establish a set of remote operations criteria that are safe, efficient, and beneficial for advanced nuclear power plants.

3.1   Remote Control Towers at Airports

London City Airport is an international airport that serves 5 million people per day. The air traffic controllers no longer work at this airport. In the summer of 2021, all air traffic control was switched to a remotely located digital control center located in Swanwick, Hampshire, which is about 100 miles away from the airport. All planes that take off and land are guided by controllers now located at the remote control center (Figure 3.1) [46].

Figure 3.1: London City Airport Remote Control Center in Swanwick [46]

Operators direct traffic using information from a 50m tall tower at the airport (Figure 3.2) [23]. The tower is equipped with 16 high-definition cameras and multiple lenses, which provide a 360-degree view of the airfield. To prevent bird damage, the tower has metal spikes on top, and each camera has a self-cleaning mechanism to prevent debris and insects from blurring the lenses. The images captured by the cameras are transmitted through multiple high-speed fiber links to a new remote control center, where they are displayed on 14 screens that offer a panoramic view of the runway [2]. In addition to the video feeds, live sound feeds are also broadcasted into the control room. The construction of the new remote tower cost $28 million, and officials assert that this isn't about saving money but rather an efficient way to expand operations. The idea was first proposed in 2016 as part of an expansion plan that required changes to the existing air traffic control tower [56]. Although the cost of building a digitized control tower may seem high, it is less expensive than rebuilding a traditional tower, which would entail the cost of constructing necessary infrastructure like roads and water supply. Moreover, the remote tower requires less maintenance and replacement of parts during its lifetime, than its analog control tower counterpart. The chief operating officer of London City Airport and his research team estimate that once the pandemic is over, City will be able to handle 45 plane movements per hour, up from 40 in 2019.

Figure 3.2: London City Airport Digital Air Traffic Control Tower

London City Airport was the first major airport to implement a remote control tower, and now other airports are also exploring this technology. Air Traffic Control Towers (ATCT) play a crucial role in directing aircraft on the ground and in airspace, ensuring safe and efficient air traffic flow. Air traffic controllers are responsible for providing pilots with updates on hazardous conditions, coordinating landing and takeoff times, directing aircraft once they've landed, and compiling data to enhance traffic and safety [50]. Typically, ATCTs are the tallest structures at airports, providing a clear 360-degree view of air traffic activities. They rely on air-to-ground communication systems such as high-frequency radio calls and visual signaling and operate under the regulations set out by the Federal Aviation Authority (FAA).

Remote air traffic tower systems, such as the one in London City, are the next generation of traditional ATCTs. Located within the airport boundaries, remote towers (RTs) utilize advanced digital technology such as distributed cameras, sensors, communication signals, and other equipment to replicate the roles of traditional ATCTs. A high-definition display provides a 360-degree view of the airport in real-time, with embedded audio signals allowing air traffic control operators to hear airport activities (Figure 3.3) [47]. Instead of 10-150 controllers in a traditional ATCT, an RT typically has three operator positions, including a ground controller, air controller, and coordinator. The RT's look may vary from airport to airport, but they all utilize a standard set of technology to fulfill their purpose.

The concept of replacing human sight from an ATCT with digital cameras in an RT is not new to the aviation industry. Since the early 2000s, digital technology has been explored to update traffic control. Technological advancements have made remote control towers more feasible, and the cost and safety benefits are driving the aviation industry to implement this technology worldwide.

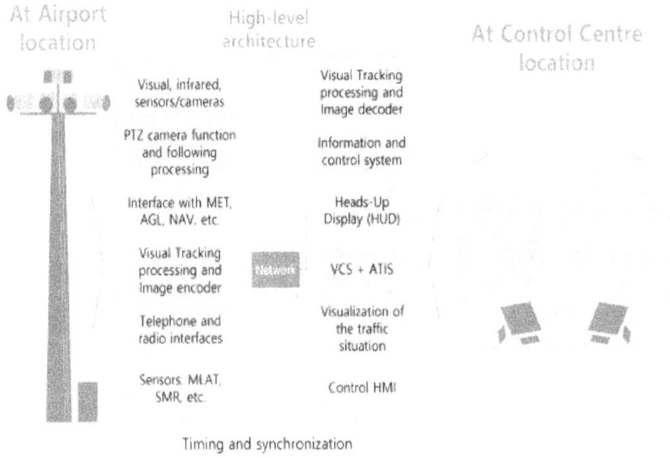

Figure 3.3: Communication Between the Digital ATCT and the Remote Control Room [55]

The utilization of technology in remote tower systems offers a unique advantage in terms of cost-effective construction of the control room. By locating the control room outside the airport, often in less expensive areas, it reduces the need for costly infrastructure development. This also means that secure communication networks and data transfer systems are crucial. Furthermore, the traditional ATCT poses a limitation to airport expansion, whereas the RT system can easily incorporate additional technology to accommodate growth.

In addition, a single remote tower module (RTM) system can service one airport, and multiple RTMs can be located in one remote tower. This enables a single control room to oversee more than one airport [47], making RTs a highly desirable long-term solution for airport growth and safety. Many airports considering expansion into the multi-airport realm or those needing to upgrade their control towers are exploring the implementation of RTs [55].

The aviation industry's adoption of remote tower systems also sets an example for the nuclear industry to follow. Both industries are subject to stringent regulatory requirements and need to address growing demand. Any technology or operational strategy changes must undergo an extensive review process. The aviation industry is successfully implementing remote control centers to cope with increasing travel demand and costs, and the nuclear industry should consider doing the same for their control rooms.

3.2   Multi-Unit Remote Control of Drones

Drones, also called unmanned aerial vehicles (UAVs), have been used by the military and civilian sectors for decades. The acronym drone stands for Dynamic Remotely Operated Navigation Equipment. Drones and UAVs are pilotless equipment operated either autonomously or from a remote-control center located on the ground [1]. The degree of remote freedom for UAVs depends on the UAV type and its function. UAVs come in slightly different forms and have military and civilian applications. To function properly, several technologies such as aerodynamics, communication, networking, and controls must be combined [20].

In civilian applications, radio frequencies are typically utilized to control drones. A remote controller sends control signals to the drone through two wireless radio frequencies of 2.4 GHz and 5.8 GHz. One frequency wave is used to transmit control signals to the drone from the ground, and the other is used to send data back to the operator [7]. Civilian drones can also utilize networks such as 4G and 5G to communicate and transfer data. In contrast, military drones were designed to monitor locations that were too hazardous or challenging for human intervention. Larger military drones are operated from a ground-based control room, and communication signals are transmitted to and from the drone via satellites, which provide widespread and consistent coverage, making them ideal for military drone use. Equipped with cameras and sensors, drones send visual and quantitative data back to the ground control station, as illustrated in Figure 3.4 [3].

Drones serve as a prime example of remote operations, featuring advanced sensors and cameras designed to perform specific tasks while being controlled from a distance. The Federal Aviation Administration (FAA) currently regulates drone usage in the United States, with regulations prohibiting an individual from controlling more than one unmanned aircraft at a time (CFR 14 part 107.35, titled "Operation of multiple small unmanned aircraft") [18]. However, the UAV industry has been exploring various operational configurations, including the innovative swarming technique, where multiple drones operate together under a single operator's control. Swarms of drones operate akin to flocks of birds or swarms of bees, working in unison to complete tasks. As the nuclear industry moves towards Small Modular Reactors (SMRs), the concept of multi-unit remote operations becomes increasingly relevant, as each SMR plant may feature multiple power-producing units controlled from a single

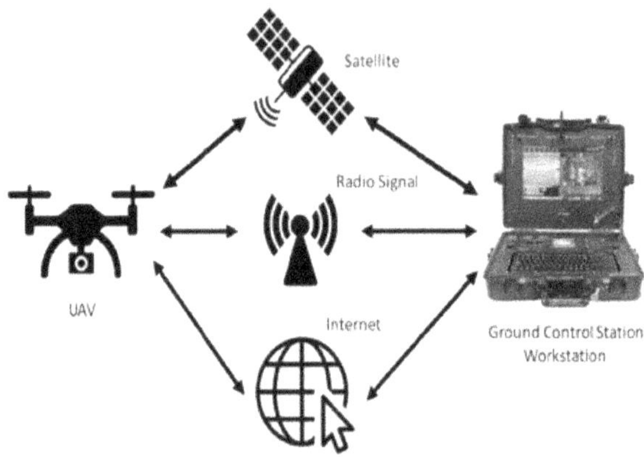

Figure 3.4: Methods of Controlling UAVs Utilizing a Ground Control Center [3]

control room. Understanding the differences between swarm operations and controlling a single drone is essential to effectively utilizing this technology.

For the swarm application, one communication infrastructure is presented in Figure 3.5 [11]. A single ground control station controls multiple UAVs that communicate utilizing radio frequencies. Another option is the FANET method (flying ad-hoc network architecture). In this instance, one UAV is connected to a satellite, and that single UAV communicates with the other UAVs using wireless communication. In this situation, the swarm tends to be more autonomous [11].

The level of autonomy for UAVs is a subject of ongoing research. Traditionally human operators monitor and control UAVs from a ground control station. It is up to the human to make decisions on behalf of the UAV. Like any other cyber-physical system, the designer must consider the most desirable autonomy level. A fully autonomous system is one in which all decisions and controls are made by algorithms, but to succeed, data must be collected from sensors, processed to produce useful information, and acted upon. The processing of data and informed decision-making is the key to autonomy. There are current examples of swarm technology with low levels of autonomy, such as coordinated light shows at events. For example, Intel deployed 300 drones to perform a coordinated light show at the 2018 Winter Olympics [28]. But to date, applications using higher levels of independence are limited.

The remote operational decisions about controlling swarms and the distribution of control between humans and machines are crucial. These decisions must consider the drones' purpose, networking capabilities, costs, and efficiency. Additionally, regulatory approval is

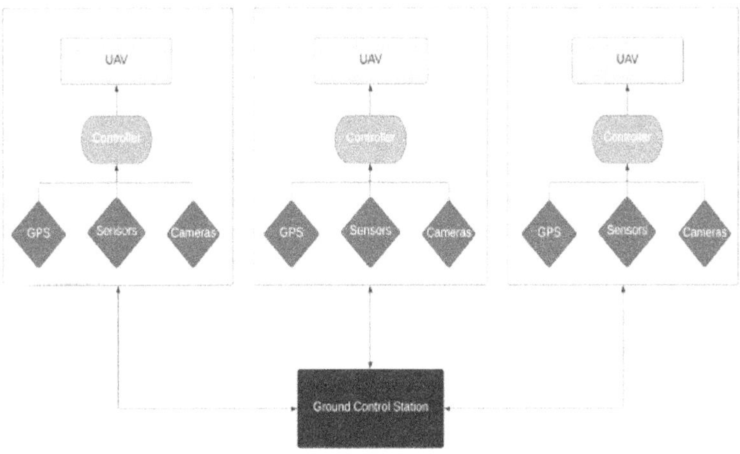

Figure 3.5: Swarm Communication Infrastructure [11]

necessary to increase the remote operations of drones. The potential benefits of remote operations in fulfilling military operations and commercial needs are significant. Operating a drone remotely can lead to obtaining information for businesses that would otherwise be unattainable or more difficult and costly—for example, inspecting leaks that are difficult to get to, health monitoring for agriculture, or inspecting offshore wind turbines and oil rigs [54]. Using drones can increase safety and reduce costs. The drone industry's transition to remote operations is ongoing, but it shows that remote operations of multiple physical systems are possible, feasible, and often more suited to the tasks at hand.

3.3 Networking of Remote Operations in the Mining Industry

Much like the other industries presented in this chapter, an increase in demand for mined commodities has led to the mining industry's work toward developing remotely operated systems. The growth in demand for mineral resources such as lithium, cobalt, and other earth metals, driven by technological advancements, has made mining a crucial industry in the manufacturing supply chain [19]. Remote operations have been developed in the mining industry to improve efficiency, productivity, and worker safety. Mining operations can be hazardous for humans, and the mines are often situated in remote and difficult-to-access locations. By creating remote operations centers, the need for workers to access such remote and potentially dangerous areas can be eliminated. This not only increases safety but also reduces costs for mining companies by eliminating the need for on-site safety features and reducing housing and other employee needs required to work on site.

The mining industry's willingness to embrace change, driven by the constant desire to optimize profitability, has led to a strong focus on IoT integration, which has already been implemented in some cases. While other industries, such as the aviation and nuclear industries, share the desire to maximize profitability, changes are still viewed as risky and are approached cautiously. Mining assets, including sensors, advanced motors, drilling rigs, switches, and gauges, are the "things" in IoT integration. The adoption of IoT, digitization, and automation across the mining supply chain creates opportunities for remote operations, data analytics, and health monitoring. The objectives of a remote operations solution in mining are to reduce employee risk, increase productivity and efficiency of employees and assets, enable operators to make better-informed decisions, provide flexibility based on changing demand for minerals, and reduce costs [44]. However, there are challenges in providing a remote operations solution for mining, particularly in terms of cybersecurity and data accessibility. As soon as a system becomes digital, cybersecurity becomes a concern, and maintaining security across the supply chain from pit to port is a priority in the mining industry. Any breach could lead to environmental incidents, resulting in fines, penalties, and potential loss of operating licenses. Despite these challenges, the mining industry's commitment to innovation and profitability means that remote operations solutions are likely to continue to be explored and implemented.

In the mining industry, network communication plays a crucial role in ensuring safety and optimizing production. To achieve this, the industry relies on various technologies, including signal processors, Programmable Logic Controllers (PLCs), and network protocols for communicating between mining assets and control rooms. To coordinate sensors, controls, and equipment, a Supervisory Control and Data Acquisition (SCADA) system is typically used for data transfer. This information is then presented and managed through interfaces within a control room [13].

As the mining industry moves towards digitization, IoT, and remote control rooms, secure and real-time data processing and access from remote locations become critical. To address this need, Cisco, a leading digital communication technology company, has developed an industrial automation solution framework that it is working to implement in partnership with players in the mining industry.

The Cisco framework offers a comprehensive communication architecture that prioritizes cybersecurity [14]. Figure 3.6 illustrates the framework, which segments mining processes both physically and logically, ensuring that issues in one area do not directly impact others. The framework features three primary segments. Firstly, the Remote Operations Center serves as the central hub, housing operators and offering complete data visualization and control across all pit-to-port operations. Communication between the ROC and the other segments occurs via an Enterprise-WAN and the internet. Secondly, the Enterprise

Zone accommodates the Enterprise-WAN, which provides connectivity between the ROC and the other segments. Lastly, the Site Operations Zone is comprised of several areas, including the Extraction Zone, which houses mining equipment, such as drills and mining vehicles; the Crushing, Processing, Smelting, and Refining Zone, which contains the physical and control infrastructure for various mine processes; the Tailings Zone, which features sensors and monitoring systems to protect workers and the mines; the Transportation Zone, which includes all transportation systems used throughout the mining chain; and the Non-Process Infrastructure Zone, which encompasses all related processes that support mine operations, such as water treatment and electrical management.

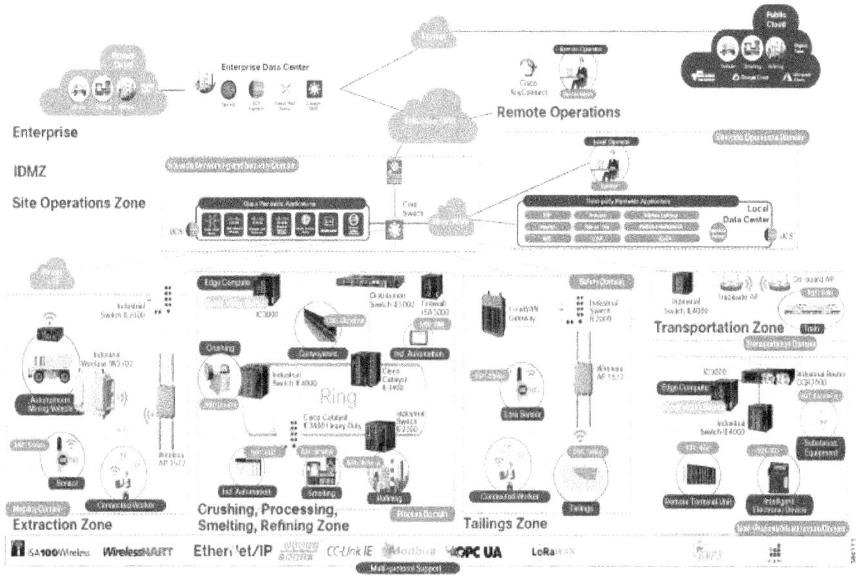

Figure 3.6: Cisco Mining Remote Operations Communication Infrastructure [14]

In the mining industry, network communication is vital for both safety and production, and a secure communication architecture is essential to meet cybersecurity goals. Cisco's framework provides just that, with a well-structured and segmented communication network that helps to minimize the impact of a problem in one area on other segments. The Remote Operations Center, Enterprise Zone, and Site Operations Zone are the three primary segments in this architecture, and wireless communication and local mine-WAN are the primary communication methods used between them (see Chapter 2 for more on networking protocols). An IDMZ, Industrial Demilitarized Zone, buffer zone acts as an additional layer of security between the trusted industrial zone and the untrusted enterprise zone, providing an extra level of protection against potential cyber threats. This is a data security strategy utilized by the Department of Homeland Security [14]. Essentially, the IDMZ is the middleman between informational technology (IT) and operational technology (OT). This helps provide additional security for the OT systems against threats coming from the IT side.

While advanced nuclear reactor power plants don't have the same distinct processes that can be segmented as those found in mining, the communication protocols and redundancy levels described in the Cisco framework can still be applied. The IDMZ buffer zone, in particular, offers an interesting security protocol that could be implemented to create an additional layer of security between IT and OT. While the evaluation of cybersecurity methods for remote operations control rooms for advanced reactors falls outside the scope of this thesis, the mining industry provides relevant models that the nuclear industry can explore.

3.4   Summary and Applications to the Nuclear Industry

In conclusion, this chapter highlighted the remote control strategies implemented in the aviation, drone, and mining industries to address various challenges related to cost, safety, scalability, and data collection. The nuclear industry can learn from these industries to optimize safety, costs, and scalability while addressing the growing energy needs and grid stability requirements. Real-time data processing, high-definition visualization, and robust communication protocols are key requirements for implementing remote operations in the nuclear industry. By adopting best practices from other industries, the nuclear industry can establish ground rules and feasibility of remote operations, paving the way for safer, more efficient, and cost-effective nuclear operations in the future.

Chapter 4

A Demonstration of Remote Operations of an Advanced Nuclear Reactor System

4.1 Background

The preceding chapters have demonstrated that the remote operation of complex systems from a remote control room is technically feasible and actively investigated and implemented in industry. Within the nuclear sector, remote operations have gained significant attention, to the extent that the U.S. Nuclear Regulatory Commission (NRC) has released a unique document titled "NRC Ground Rules for Regulatory Feasibility of Remote Operations of Nuclear Power Plants" [60]. Unlike most NRC regulatory documents, this official memo has a flexible and exploratory tone. It begins with a disclaimer that acknowledges ongoing discussions and specifies that the contents are subject to change, indicating the document's uncertain nature:

"This report has been prepared and is being released to support ongoing public discussions. This report has not been subject to NRC management and legal reviews and approvals, and its contents are subject to change and should not be interpreted as official agency positions."

It is not unexpected to encounter non-committal expressions in the context of remote operations in the nuclear industry, given that this is a relatively new area. Currently, there are no guidelines available that specifically address this operational approach. Existing industry regulations related to reactor control, including operator licensing and control decisions, pertain solely to on-site control rooms located within the reactor site boundary.

The authors of the NRC document describe it as a preliminary identification of items and considerations that need to be explored and addressed by anyone interested in pursuing the remote operations control strategy. The criteria relevant to this dissertation are summarized here [60]:

Ground Rule #1 – Remote operations must be part of developing a nuclear reactor from the outset. Remote operations cannot be retrofitted or added to the design.

Ground Rule #3 – Regulatory changes may be necessary to accommodate remote operations.

Ground Rule #6 – Data, communication, and security infrastructure is critical and must be conceptualized early in the design of the control room.

Ground Rule #7 – The responsibilities of the remote operators will need to be determined based on automation and "minimal risk conditions." Identified responsibilities should support decisions about the number of controlled facilities, operators, and operator training.

Ground Rule #8 – Training and licensing requirements for Operators will need to be determined.

Ground Rule #9 – Having a crew based on site that can take control in case of operational issues may be unavoidable.

These ground rules outlined above are broad in scope, which is understandable given the limited knowledge about remote operations in other industries and the fact that the remote operations paradigm has never been explored before in the nuclear industry. These rules offer advanced reactor companies some degree of flexibility in designing a novel type of control room and collaborating with regulatory bodies, rather than adhering to strict guidelines. This freedom can be both thrilling and daunting. Nonetheless, it is crucial to note that several ground rules would have needed consideration whether the control room was onsite or offsite. For the first time in the nuclear industry's history, control rooms are designed to be fully digitized and contain some level of automation from the outset. Implementing a digital control room necessitates establishing a networking protocol between the physical system and the control room while considering cyber security defenses. With the digital control room and automation, the roles of operators change, and operator training will need to reflect those changes. The number of units the digital control room can control will need to be decided. Furthermore, the development of passively safe reactor designs, where systems are actuated by removing electrical power, simplifies the safety-related protective functions, which can remain local while plant control and health monitoring are transferred to a remote location. Finally, regulations must consider the new ways in which operators interact with digitization, automation, and the overall new plant designs. Thus, the majority of the questions that advanced reactor companies must address during the design process do not differ substantially between a digital local control room and a digital remote control room. However, the importance of addressing each issue may be greater in the latter case.

Both the NRC and the iterative design process emphasize that a remote control room cannot be retrofitted to an advanced nuclear reactor plant design, nor should it be considered late in the design and development process. With advanced reactor companies like TerraPower, X-energy, and Kairos Power aiming to deploy their designs within 15 years, it is critical to commence the iterative design process promptly if the remote control room is to be included in the deployed designs. While all three nuclear reactor companies mentioned are contemplating some form of a remote control room, there is limited information available on what these rooms might look like due to proprietary reasons or a lack of designs. It appears that the remote control room is still in its initial ideation stage.

The following chapter will demonstrate the technical feasibility of remote operations for the nuclear industry. This chapter presents a documented demonstration of a remote control room capable of controlling an initial test loop of the advanced reactor concept of a fluoride salt-cooled high-temperature reactor or an FHR. This author worked with the advanced reactor company, Kairos Power, to control their major test facility, the Engineering Test Unit (ETU) in Albuquerque, New Mexico, from Kairos Power headquarters in Alameda, California.

4.2   Study Demonstration Design and Method

Purpose

The objective of this study is to showcase the technical feasibility of remote operations for the Kairos Power ETU using a control room concept of operations that can serve as a representative model for any forthcoming KP-FHR control room. The control room concept of operations pertains to the operator's roles and responsibilities, the operational environment, and how the users interact with each other and the system. By conducting a documented demonstration, this author aims to encourage advanced reactor companies and the NRC to refine the remote control room design through iterative improvement based on this initial benchmarking.

Exclusions and Limitations

This study will not consider any regulatory considerations for remote operations, control room design, or operational strategies. As an electrically heated reactor prototype, the ETU does not have any nuclear safety functions. The objective here is to focus solely on the technical feasibility of remote operations. This study explored some limited aspects of cyber security. However, this study did not test or evaluate the level of cyber security currently built into the data/communication infrastructure. Furthermore, the specific HMI designs and networking protocols used are proprietary information and cannot be disclosed within this dissertation. However, all information needed to demonstrate technical feasibility for remote operations will be discussed.

Operators and Scenarios

For this test, two operators were in the remote control room, and two were in the local control room. While not officially licensed operators, all operators were individuals with extensive knowledge of the ETU system and the authority to control the ETU system. At Kairos Power, operators are referred to as either Test Engineers (TE) or Test Directors (TD). The TE performs all control actuation as directed by the TD and based on procedures. At the start of a procedure, the TD will signal to the TE that they can proceed through the steps, and the TE communicates all actions taken to the TD once a step is completed. The TD will supervise all steps taken from their workstation and inform the TE when they can take the following action. The TD monitors all TE actions and may take necessary measures to ensure safe operations. The Test Engineer role is equivalent to a conventional Reactor Operator, while the Test Director role is similar to that of a Supervisor who oversees the overall system. For this demonstration, the following operators were present:

- 1 Remote TE (RTE)

- 1 Remote TD (RTD)

- 1 Local TE (LTE)

- 1 Local TD (RTD)

In this demonstration, the Reactor Control and Shutdown System (RCSS) was tested. The RCSS is Kairos' first iteration of the control rod system utilized by nuclear reactors to regulate the rate of fission reactions. A control rod system is a critical part of the design of any advanced reactor, as it is one of the systems responsible for maintaining reactor safety. Given its relative importance to a nuclear system, this author determined that demonstrated remote control of the RCSS would be valuable. All command and control during the duration of this demonstration came from the control room in California. For this scenario, the electricity supply for the reactor control shutdown element was powered on locally, then fully powered on by the remote control room operators. Next, the operators manually withdrew and inserted the control elements or initiated the automatic cycling functionality built into the HMI. Finally, the RCSS system was powered off to conclude the scenario. The scenario steps are detailed below:

1. Verify a communication link is established between the remote and local control rooms.

2. The remote control room verifies with the on-site test engineers that all manually actuated argon supply and isolation valves are in their correct positions for testing.

3. The remote control room verifies with the on-site test engineers that the Electrical supply for the RCSS system is powered on.

4. The remote and local control rooms verify HMI signals by ensuring pressure readings, thermocouple readings, and insertion percentages are within the expected ranges for the startup of the system.

5. RTE engages the RCS clutch.

6. RTE withdraws the RCS rod using the HMI soft control until the avg position is 55%. Then uses soft controls to stop the rod movement.

7. RTE withdraws the RCS rod using the HMI soft control until the avg position is 10%. Then uses soft controls to stop the rod movement.

8. RTE inserts the RCS rod using the HMI soft control until the avg position is 90%. Then uses soft controls to stop the rod movement.

9. RTE uses soft controls to start auto cycle functionality, which initiates automatic control rod travel.

10. RTE and RTD monitor auto travel for 3 complete withdrawal and insertion cycles.

11. RTE enters a percent insertion value of 70% and a percent withdrawal value 0f 20%.

12. RTE and RTD monitor auto travel for 2 complete withdrawal and insertion cycles.

13. RTE attempts to enter a percent insertion value that is out of bounds, to test the HMI threshold limits.

14. RTE attempts to enter a percent withdrawal value that is out of bounds, to test the HMI threshold limits.

15. RTE turns off the auto travel functionality using the soft controls.

16. RTE sets the insertion value to 100%.

17. RTE and RTS monitor the rod insertion until it has reached 100% insertion and verify the rod is holding in that position.

18. RTE release the clutch using the clutch soft control.

19. RTD communicates to the local control room that the test is complete.

20. The local control room took back command and control of the RCSS system by removing control capabilities from the remote control room. This final action serves as an initial test of some cyber-security functionality.

ETU and the Local Control Room

The Kairos Power Engineering Test Unit is a non-nuclear, unenriched Flibe-wetted integrated test. The Kairos Power FHR reactor or the KP-FHR will use Flibe, a fluoride lithium beryllium salt, as a coolant. This coolant's high heat capacity and high boiling temperature make it suitable for high-temperature operating conditions. The KP-FHR will use TRISO pebbles, tri-structural isotropic particle fuel, which are carbon and ceramic-based kernels filled with uranium fuel. The ETU is the first integrated iteration of the KP-FHR design. It contains a vessel, pump, pebble handling system, a control element shutdown system, and other sub-systems necessary for the functioning of the KP-FHR, a digital local control room, and a digital remote control room. The purpose of the ETU is to demonstrate KP-FHR technologies while gaining valuable insight for improvements as Kairos continues to iterate and scale up its designs [33].

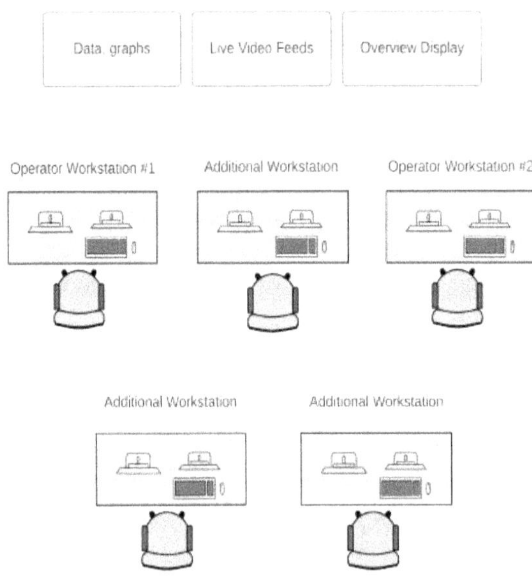

Figure 4.1: ETU Control Room Layout located at KP-SW in Albuquerque, New Mexico

The local control room is located at KP-SW in New Mexico. The control room contains five workstations, all equipped with two 4k displays per workstation. There are three large monitors located in the front of the control room (Figure 4.1). One monitor is used as an overview display, and the other two contain live video feeds of the ETU enclosure. While it is unlikely that a true nuclear power plant control room would utilize five workstations, maintaining this setup has proved to be beneficial during the initial testing phases. Typically, during operations, two workstations are used by the two operators performing tests. The other workstations can be used to run the KP simulator or for stakeholders to observe aspects of the ETU as tests are happening. Controls for the ETU are digital, and the HMIs were developed utilizing an industrial software platform. All data from the ETU is displayed on the HMIs, and all control actuation comes through the HMIs. In addition, a server-client communication architecture is utilized. Data from the ETU is collected within PLCs. PLCs are connected to a server, which sends the real-time data to the operator workstation computer, which acts as a client. A general communication architecture is presented in Figure 4.2.Data is passed between the PLC, Server, and clients utilizing OPC UA (see Chapter 2 for details). The red line indicates the OPC UA connection. Additional details about the HMI development and ETU networking protocol are proprietary.

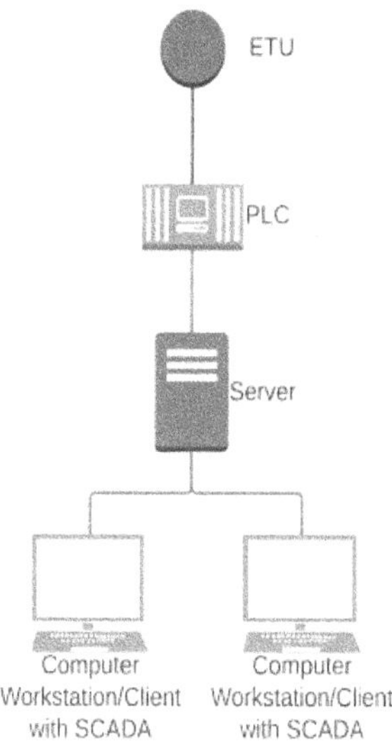

Figure 4.2: ETU to Control Room General Communication Architecture

Remote Control Room Environment

The remote control room is located at KP-HQ in California. The remote control room has two operator workstations with two 4k displays. In addition, three large monitors are located in the front of the room. At each operator workstation is an RTS audio panel that allows for direct audio communication to the ETU local control room at KP-SW. The layout is presented in Figure 4.3.

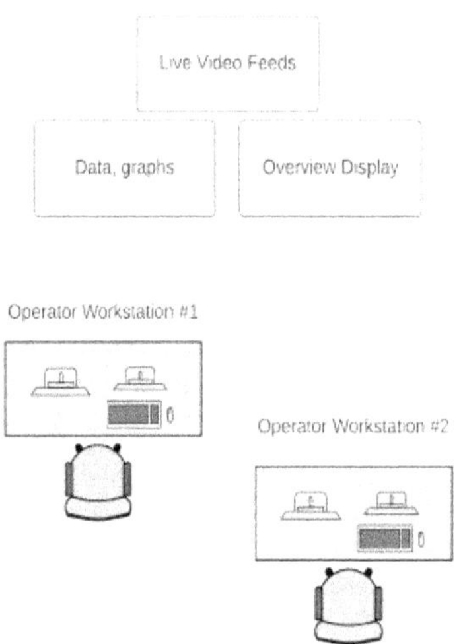

Figure 4.3: Remote Control Room Setup Located at KP-HQ in Alameda, California

The remote control room utilizes VPN and IP addresses to access the networking protocol at the local control room. Cyber-security measures are in place but remain proprietary and cannot be disclosed within this dissertation. Additional engineering controls and administrative protocols enable user identification, location tracking, and user access approval, with the ability to revoke access as needed. For the day-to-day operations of the ETU, the remote control room is a backup control room. It has never been used to run tests with the ETU system beyond testing general HMI functionality.

4.3   General Study Protocol

In this study, two operators worked together to remotely operate the ETU facility at KP-SW in Albuquerque, New Mexico. The control over the ETU came from the control room center at KP-HQ in Alameda, California. During the study, the operators utilized the HMIs located at the workstations in the remote control room and worked through an ETU procedure that controls the control rod shut-down system for the ETU. Two operators were on standby in the local control room for the duration of the demonstration. There was clear and consistent communication between the two control rooms using Zoom and the RTS audio panel. The local control room was instructed to take over command and control should any issues arise. It is worth noting that the HMIs and controls in both the remote and local control rooms were identical. Both control rooms had access to the same information and had the same ability to control the ETU. Therefore, there was no technical difference between the control rooms, except for their physical distance from the ETU.

This study aims to serve as a small-scale demonstration that remote operations of an advanced reactor is technically possible. From this baseline, Kairos and other advanced reactor companies can iterate on the remote-operations control room design and the operational strategy for remote operations of future systems. Additionally, regulatory feasibility can be considered after this initial baselining of the technical feasibility. To measure the success of this demonstration, the following metrics were decided upon:

1. Remote Operators complete the entire procedure without local control room intervention.

2. Clear communication is maintained between all operators throughout the procedure.

3. Successful control of the ETU is achieved using the Ignition HMIs.

Training

Before starting the study, all operators were familiarized with the HMIs and procedures. The operators were given their roles and the procedure ahead of time. To remove uncertainty, all operators needed to gain familiarity with the control room, the control systems, and the communication protocols and ask any questions before the demonstration. Therefore, the remote TE and TD had access to the HMIs and the procedure. There were several working sessions between the remote and local control rooms before the demonstration to refine the HMI and controls and ensure data communication and communication between the two control rooms worked well.

Survey

At the end of the procedure, the remote TE and TD were asked the following questions:

- On a scale of 1-5, 1 being the lowest and five being the highest, how secure did you feel while performing the procedure? Please explain your ranking.

- On a scale of 1-5, 1 being the lowest and five being the highest, how well did you understand what was happening with the RCSS system throughout the procedure? Please explain your ranking.

- On a scale of 1-5, 1 being the lowest and five being the highest, how successful were you communicating between the remote and local control room? Please explain your ranking.

- Did being in a remote space away from the system you control impact how you felt about the tasks you were asked to do?

- Did you trust the ability to control ETU from a remote-control room? Please explain.

The purpose of these questions was to gauge the operators' comfort levels and understanding while remotely operating a system that was not physically nearby. Through discussions with various stakeholders at the advanced reactor company and within the nuclear industry at large, it has become apparent that there exists a mental block when it comes to remote operations (which will be discussed in greater detail in Chapter 6). The notion of remote control makes some people apprehensive. This dissertation posits that with proper design of the control room, including the HMI design and OSSs, and with the nuclear system designed to minimize risks including the separation of protection systems from control systems, the geographic location of the control room need not have a negative impact on plant operations. The questions presented above were intended to provide some initial data on how in-tune the remote operators were with the system, based on the initial design concept of the control room, and to offer feedback on improving the design of the remote control room.

4.4   Results and Discussion

The remote operators successfully controlled the RCS system and achieved all three success metrics outlined in section 4.3. They used the HMI to manage the RCS system and completed the entire procedure without any intervention from the local control room. The control rod movement throughout the procedure is depicted in the Figure 4.4, with the green and blue plot lines representing two real-time data measures monitored throughout the procedure. The y-axis reflects the % insertion, with a value of 100 signifying full insertion, while the x-axis depicts real-time. The plot illustrates the numerous cycles the RCS underwent, culminating in the rod being fully inserted to 100%. Zoom proved to be the primary mode of communication between the local and remote control rooms during this initial demonstration, with a video displayed on TV monitors in both locations. Alternative communication methods may be explored during subsequent control room iterations. Finally, at the end of the procedure, the local control room utilized an initial cybersecurity action incorporated into the HMI and terminated the access of the remote control workstations to the entire Ignition system.

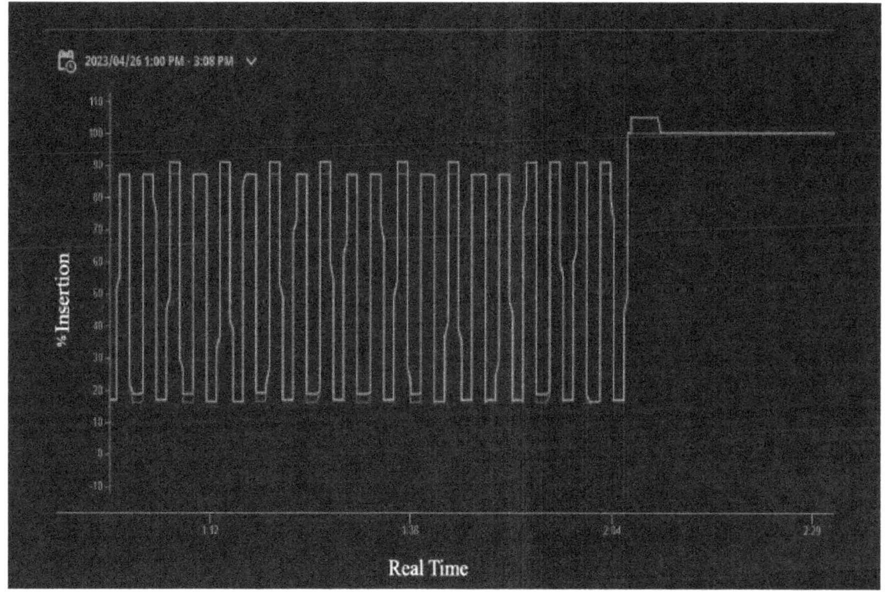

Figure 4.4: RCS Cycles Plot

Table 4.1 and Table 4.2 presents a summary of survey question responses obtained from the Remote Test Director (RTD) and Remote Test Engineer (RTE) involved in the remote operations study. The results indicate that the overall experience of the operators was positive, as they felt comfortable operating the HMI and successfully completed the procedure without any intervention from the local control room. The availability of real-time data proved to be a valuable tool, enabling operators to make informed decisions throughout the process. Additionally, the Zoom platform served as an effective means of communication between the remote and local control rooms, enhancing the overall efficiency of the operation.

However, the RTD expressed some concern regarding a delayed response from the HMI control action to the control signal in the hardware. Nevertheless, as the RTD observed, the real-time data revealed that the same delay was observed in the local control room, indicating that the issue was not related to remote operations, but rather a programming logic error within the PLCs. Overall, the results of the survey indicate a high level of satisfaction among the operators, providing further evidence of the feasibility of remote operations for advanced nuclear reactor systems.

4.5   Conclusion

In conclusion, the successful demonstration of remote control for an iteration of a future nuclear system marks a significant step towards realizing the full potential of remote operations in the nuclear industry. This demonstration showcased the ability of remote operators to manage the RCS system and achieve all success metrics without the need for intervention from the local control room. The use of HMI and real-time data monitoring proved to be effective tools in enabling remote operators to manage the RCS system, while the Zoom platform facilitated seamless communication between the remote and local control rooms. The incorporation of cybersecurity measures into the HMI also highlights the importance of addressing security concerns associated with remote operations.

The success of this demonstration provides a strong foundation for future research and development of remote operations in the nuclear industry. The results of this study can be used to guide the design of future control room iterations and to identify areas for improvement. Further research can also explore the feasibility of remote operations for other aspects of nuclear power plant operations, such as maintenance and repair activities. Ultimately, remote operations have the potential to enhance the safety, efficiency, and flexibility of nuclear power plant operations, and the successful demonstration of remote control for an iteration of a future nuclear system represents a crucial step towards realizing this vision.

Table 4.1: Remote Operations Demonstration Survey Question Responses from the RTD

Survey Question	RTD Response
On a scale of 1-5 (1 = low, 5 = high), how secure did you feel while performing the procedure? Please explain your ranking.	4 - I felt very secure while performing the procedure knowing that the remote control room has limited access
On a scale of 1-5 (1 = low, 5 = high), how well did you understand what was happening with the RCSS system throughout the procedure? Please explain your ranking.	4 – I understood what was occurring with the RCS throughout the procedure
On a scale of 1-5 (1 = low, 5 = high), how successful were you communicating between the remote and local control room? Please explain your ranking.	4 - I believe the communication was successful
Did being in a remote space away from the system you control impact how you felt about the tasks you were asked to do?	No, I was still nervous and hopeful that everything would go smoothly and felt responsible for my actions
Did you trust the ability to control ETU from a remote-control room? Please explain.	I trusted the ability to control the ETU from the remote control room with my only concern being the delayed response of the system to operator inputs. However, I believe this delay is present in the local control room and is more of a system/software delay and not so much a remote operations delay.

Table 4.2: Remote Operations Demonstration Survey Question Responses from the RTE

Survey Question	RTE Response
On a scale of 1-5 (1 = low, 5 = high), how secure did you feel while performing the procedure? Please explain your ranking.	5 - The procedures were clear enough for us to understand. Knowing that the local ops team could take control at any moment was reassuring. Not because they were closer to the system, but they were more familiar
On a scale of 1-5 (1 = low, 5 = high), how well did you understand what was happening with the RCSS system throughout the procedure? Please explain your ranking.	4 - The only thing that wasn't clear was if 100% meant fully inserted or fully withdrawn
On a scale of 1-5 (1 = low, 5 = high), how successful were you communicating between the remote and local control room? Please explain your ranking.	5 - I thought communication over Zoom was fine
Did being in a remote space away from the system you control impact how you felt about the tasks you were asked to do?	Not really, being in the local control room vs the remote control room feels very similar since the screens are identical
Did you trust the ability to control ETU from a remote-control room? Please explain.	Yes, the HMI gave good feedback on what was happening and gave the same amount of feedback that you would get in the local control room.

Chapter 5

Expanding Remote Operations: Evaluating the Feasibility of Controlling Multiple Plants from a Single Control Room through a Human Factors Study

5.1 Background

Renewable energy sources coupled with the geographic distribution of power plants (gas, or nuclear etc.), have created challenges for the electric grid to manage. As energy consumption increases and fluctuating energy sources become more prevalent, flexible and efficient power generation will become critical. SMR designs are great candidates for providing flexible and efficient generation. The modular nature of these reactors means that a utility or a state government can purchase as many modules are needed to produce a specific amount of electricity.

As a result, nuclear power plant operations will expand into the multi-unit realm, where one control room will likely be responsible for several reactor modules. Furthermore, it is possible a single control center could be used to control several nuclear power plant sites, each with several reactor modules. Another aspect of the SMR design is their ability to load follow, adjusting the amount of power coming from an NPP based on grid demand. There are cases of NPP load following in other countries, such as France, but this option for NPPs has yet to be widely used within the U.S [29]. The increasing instability of the grid due to fluctuating renewable energy output, increasing demand from the grid, and the resurgence of the nuclear industry together open the opportunity for gen IV reactors to load follow. Load-following capabilities paired with remote operations create a powerful advantage where several NPPs can be controlled from a single off-site control room, where the NPPs are beholden to changing grid demand, providing energy security for the nation's near-term energy needs.

A single remote control room that controls several plants that load follow and provide base power is a desirable end goal for the NRC and advanced reactor companies. However, a gap exists in our knowledge of how the operations of multi-units envisioned for SMRs will work. There is also little known about controlling multiple plants, each with multiple units, from a single control room located far away from the reactors the control room controls. Therefore, studying how operations of multi-unit plants from a remote control center may play out is of extreme relevance and importance.

Several factors need to be considered for control room design and control operations of a multi-unit, multi-plant remote control center. The interface design is critical and must be able to communicate to the operators all information from each unit and plant, and provide controls that enhance situational awareness and remove confusion. How many operators are needed? How tasks are distributed amongst the operators is another essential question. Does one operator control one plant, meaning they would control all the units within a single plant? Is there one senior reactor operator to monitor several plants, or is there one SRO per plant? What are the roles and responsibilities of each operator? The answer to these questions are likely dependent on how many NPPs are being controlled from the control room, how many units per NPP, how many operator crews are needed to maintain the safety of all plants, and the technical characteristics of the nuclear power plants. The safety systems of the NPP and the level of automation will influence the answer to some of these operational questions. The number of operators must strike a balance between maintaining the security of all NPPs, coordinating grid demand, and keeping operations and maintenance costs reasonable. The roles and responsibilities of the reactor operators and senior reactor operators will depend not only on the number of units and plants but also on the design of the reactor and its controls. As discussed in earlier chapters, the safety features of SMRs and the control automation will change how operators interact with the reactors they control. It is conceivable that the specific physical safety features of the plants and the controls design will play a prominent role in determining how many operators are needed, their roles, and the number of units, plants, and control rooms they must oversee. The following chapter describes a human factors based study baselining an operational strategy for SMRs that tests the control of two plants each with four units, with 3 operators, from a single control room.

Traditional Operator Roles and Responsibilities

Traditional LWRs in the U.S. maintained crews of a minimum of 4 operators per unit. Below are definitions and descriptions of operating crews and operator roles [58, 31]:

- An **Operating Crew** is a group of personnel that works in the reactor control room on a shift basis.

- A **Reactor Operator (RO)** has historically overseen controlling the nuclear reactor. They are responsible for moving control rods, turning on/off, and stop/start for appropriate equipment. The RO makes all control actions required based on procedures.

- A **Senior Reactor Operator (SRO)** can perform the operator duties of an RO but are licensed to supervise and direct the reactor operators in their activities.

- The **Senior Technical Advisor (STA)** is an operator that supports the SRO during transients and accident scenarios.

- The **Shift Manager** is responsible for all activities during each shift. In addition, they ensure compliance with safety protocols and procedures. Control room supervisors report to the shift manager.

Number of nuclear power units operating[a]	Position	One Unit	Two units		Three units	
		One control room	One control room	Two control rooms	Two control rooms	Three control rooms
None	Senior Operator	1	1	1	1	1
	Operator	1	2	2	3	3
One	Senior Operator	2	2	2	2	2
	Operator	2	3	3	4	4
Two	Senior Operator		2	3	3	3
	Operator		3	4	5	5
Three	Senior Operator			3	4	
	Operator				5	6

Figure 5.1: Minimum Requirements Per Shift for On-Site Staffing of Nuclear Power Units Under 10 CFR Part 55 [61]

As seen in Figure 5.1 from NRC guidelines, the number of operators and control rooms increases with the number of operating units. Historically, the NRC has dictated in 10 CFR part 50.54(m) that a one-unit NPP should have two Senior Operators, two Reactor Operators, and two Non-licensed Operators. As the number of units per plant increases, the number of the Reactor Operator and Non-licensed Operator increases by 1. Therefore, a three-unit NPP would need two Senior Operators, four Reactor Operators, and four Non-Licensed Operators. Additionally, there would need to be two control rooms for the three units [61]. However, the NRC rules applied to the previous generation of LWR reactors. Advanced Reactor companies are working with the NRC to adjust these rules based on the designs of their reactors. Safety features, digitization, and automation substantially reduce the number of operators required to control a gen IV reactor. Some studies have been conducted that utilize only 2-3 operators to control a representative single unit advanced NPP [53].

NuScale, the first SMR company to detail its multi-unit operating plans to the NRC, brought its control room crew down from six operators to three running multiple units. The company's staffing plan details are presented in an NRC Topical Report. The NuScale plant is designed with 12 units controlled by a single control room, with two SROs and one RO. The STA position was allowed to be removed as a stand-alone position and instead was merged with one of the SRO positions. The allowance to combine these roles was primarily due to the development of advanced human-system interface upgrades [49]. The HMI support tools replace the advice and, in some cases, the control actuation that would have come from the STA in transients or accident scenarios. While NuScale does not provide the public with specifics on their operators' exact roles and responsibilities, the RO likely maintains the responsibility of the control actuation for up to 12 units, and the SRO directs operational activities. The second SRO acts as a shift manager that is aware of all operations and maintenance activities and ensures compliance with safety protocols and procedures.

From the NRC remote operations ground rules document described in the previous chapter, this chapter focuses on one of them [60]:

Ground Rule #7 – "The responsibilities of the remote operators will need to be determined based on automation and 'minimal risk conditions.' Identified responsibilities should support decisions about the number of controlled facilities, operators, and operator training."

One of the many advantages of remote operations is the potential to control not just multiple units within a single plant, but also the ability to oversee several plants situated in different geographic locations from a single control room. However, determining the number of facilities, operators, and their respective roles and responsibilities remains an open question. This study aims to build on NuScale's operational strategy by investigating whether three operators can manage two distinct plants with multiple units from a single control room without experiencing a significant increase in mental workload. To this end, this author developed a human factors-based study that will be described in the following sections. What follows is a description of the study laid out in the context of a concept of operations intended to represent a plausible ConOps for the remote operations of advanced reactors in the future.

5.2　Concept of Operations

A concept of operations, or ConOps, is a user-oriented document that communicates the characteristics of a system. It describes your system's who, what, when, where, and how. In this case, a ConOps was developed for one or two advanced reactor plants operated from a single control room. The ConOps is presented below, and the study methods are woven into each section of the ConOps.

"Who" - Operators and Participants

The "who" are the operators. The operators, comprising a crew of three, are the key players in this ConOps. Two of them will be designated as ROs and will be responsible for executing procedures and control actuations. The third crew member will take on the roles of both the SRO and the STA. This individual will supervise the control actions, monitor maintenance schedules, and offer assistance to the ROs as necessary. The operator groups are divided into two categories. The first group of three will manage a single plant with four units from a single control room. The second group of three will have RO 1 controlling plant 1's four units while RO 2 manages plant 2's four units, with one SRO overseeing both plants. Their respective roles are further elaborated below:

Category 1: Single plant, four units, three operators, one control room

- Reactor Operators (RO): will be responsible for any control actuation for up to 4 units. They are responsible for following the applicable procedures and directly interacting with the HMI to control the plant. They report directly to the SRO and must frequently update the SRO on the actions they are carrying out. If an issue arises, they must inform the SRO and work together to determine the best course of action. The ROs can decide how to divide up control of the four units and must notify the SRO of their control strategy.

- Senior Reactor Operator (SRO): will supervise activities at the plant. They will oversee monitoring the overall status of the plants, ensuring the correct procedures are being followed at both plants, and maintain awareness of plant efficiency and revenue coming from both plants. The SRO will also alert ROs to maintenance and accordingly direct plant activities. If needed, the SRO can step in as an RO and help ROs make decisions during abnormal operations.

Category 2: Two plants, four units each, three operators, one control room

- Reactor Operator (RO) Plant 1: will be responsible for any control actuation for up to four units that pertain to plant 1. They are responsible for following the applicable procedures and directly interacting with the HMI to control the plant. They report directly to the SRO and must frequently update the SRO on the actions they are carrying out. If an issue arises, they must inform the SRO and work with them to determine the best course of action.

- Reactor Operator (RO) Plant 2: will be responsible for any control actuation for up to four units that pertain to plant 2. They are responsible for following the applicable procedures and directly interacting with the HMI to control the plant. They report directly to the SRO and must frequently update the SRO on the actions they are carrying out. If an issue arises, they must inform the SRO and work with them to determine the best course of action.

- Senior Reactor Operator (SRO): will supervise activities at Plants 1 and 2. They will oversee monitoring the overall status of the plants, ensuring the correct procedures are being followed at both plants, and maintain awareness of plant efficiency and revenue coming from both plants. The SRO will also alert ROs to maintenance and accordingly direct plant activities. If needed, the SRO can step in as an RO and help ROs make decisions during abnormal operations.

Participants

The SRO requires a higher level of knowledge and skill than an RO, for this reason the study lead assumed the role of the SRO for all groups during this study. Therefore, two participants were needed for each study trial, and each would act as an RO.

The study participants were from the UC Berkeley Nuclear Engineering Department and a local advanced reactor power plant company. While non-experts could operate Rancor, the participants' knowledge of nuclear systems was a valuable asset for this study, as it reduced the learning curve for operating Rancor and minimized the underlying stress of working with unfamiliar territory. All participants were between 18 and 50 years old, and the participant pool was gender-diverse. Although the participants had little to no experience with control operations, they were randomly assigned roles as ROs, RO 1, or RO 2. With a total of 36 participants, the study was conducted with 18 teams, who worked collaboratively to complete scenarios, maintain plant safety, and load follow.

"What" - Rancor Simulator

The "what" refers to the apparatus that the users are interacting with. This study utilized a simulator developed by Idaho National Laboratory (INL), Rancor, to simulate a multi-unit nuclear power plant. Rancor is described as a microworld, a simplified simulator that reproduces certain critical physical phenomena while allowing for the manipulation of controls. Microworlds are beneficial because they enable very complex systems to be stripped down to their fundamental dynamics and can be more easily understood and controlled by participants that lack extensive knowledge about real-world systems [37]. Utilizing non-expert participants is highly valuable when conducting human factors experiments. This is because unfortunately, human factors-based studies are often plagued with limited participant pools, and within the nuclear industry, participation has historically been low, given the finite number of individuals with operational expertise. Utilizing the Rancor microworld allows for student participants to test theoretical concepts of controlling a reactor.

The Rancor microworld simulates a simplified water-based nuclear power plant that contains the nuclear reactor core and the pumps, pipes, and loops that make up the basics of a nuclear power plant Rankine steam cycle. In the primary loop, reactor coolant pumps are controlled to increase flow, and control rods can be adjusted to change reactivity in the core. On the secondary side, operators can control feedwater pumps and control flow to the steam generators, which then feed into a turbine. The basic physics of the primary and secondary sides is programmed into the simulator, and it is up to the operators to manually control the plant with alarms programmed into an alarm panel and interlocks that aim to keep the plant within safe parameters. Figure 5.2 is the original Rancor user interface design (P&ID), with the operator control options located at the bottom of the interface and the alarm panel at the top [37].

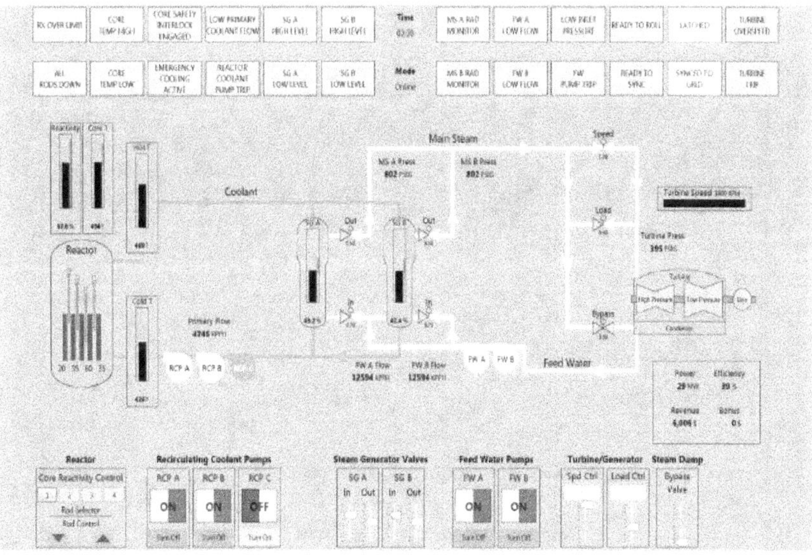

Figure 5.2: Single Unit Rancor Simulator [37]

For this study, Rancor was updated and expanded with multi-unit capabilities and the ability to automatically adjust power output based on grid demand. In the multi-unit version of Rancor, participants can control up to four units using either manual or automatic modes. In manual mode, the participants manually adjust control rod positions, recirculating pumps, steam generator valves, feedwater pumps, and turbine control for each unit. In contrast, auto mode allows participants to input a target power output percentage for each unit, after which the Rancor simulator automatically adjusts the pumps, steam generators, valves, and turbines to achieve the set power output. This automation reflects the current trend towards more automated control strategies in the industry, making the auto mode more realistic and representative of real-world operations than manual mode, and is therefore used in this study.

The multi-unit version of Rancor provides participants with information on demand, efficiency, and revenue. Demand refers to the incoming grid demand, efficiency indicates how efficiently the reactor is running, and revenue shows how much revenue the entire plant generates at any given time. All the data generated during each session, including demand, efficiency percentage, and revenue, are time-stamped and saved by Rancor, providing valuable data points for analysis.

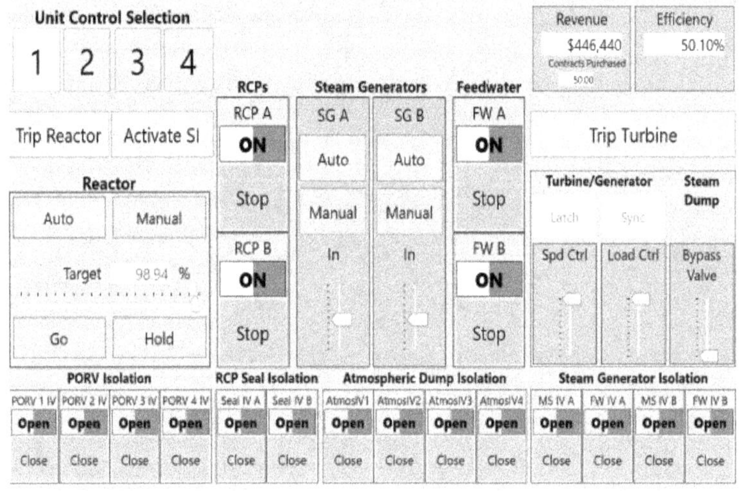

Figure 5.3: Updated Multi-unit Rancor Controls

As advanced nuclear power plants (NPPs) will likely have multiple human-machine interfaces in their control rooms, it is important that each HMI presents unique information to the operators to avoid information overload. In order to address this issue, Rancor has three separate HMIs. The first HMI, shown in Figure 5.3, is the control HMI and serves as the primary interface for controlling all Rancor units. On the other hand, the second HMI, shown in Figure 5.4, is the P&ID of the Rancor plant. The Rancor P&ID provides a visual representation of the nuclear power plant and its various systems, including the

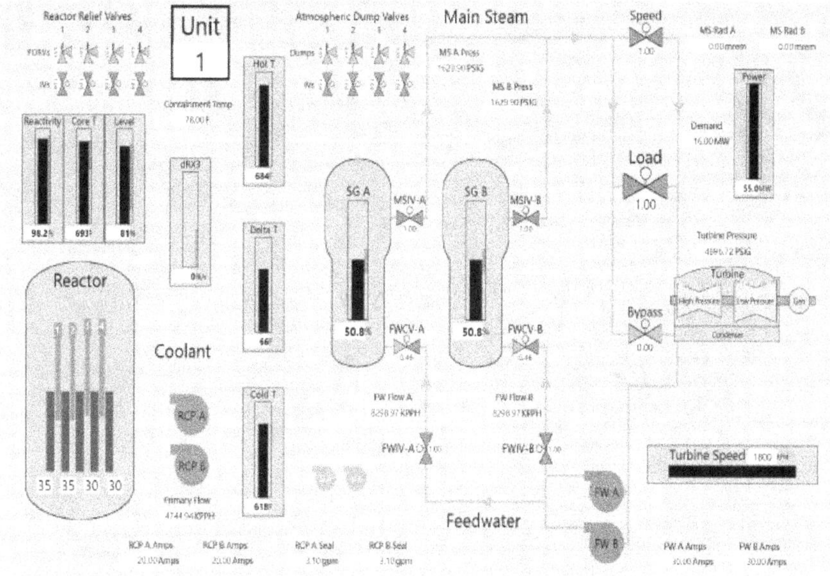

Figure 5.4: Updated Multi-unit Rancor P&ID

primary and secondary cooling systems, reactor coolant pumps, steam generators, turbines, and associated valves and controls. The P&ID is an important reference tool for operators, providing them with a clear understanding of the layout and interconnections of the various systems within the power plant. The third HMI displays each unit's alarm panels and critical values (Figure 5.5). This supervisory screen is designed to provide operators with essential information about each unit at all times.

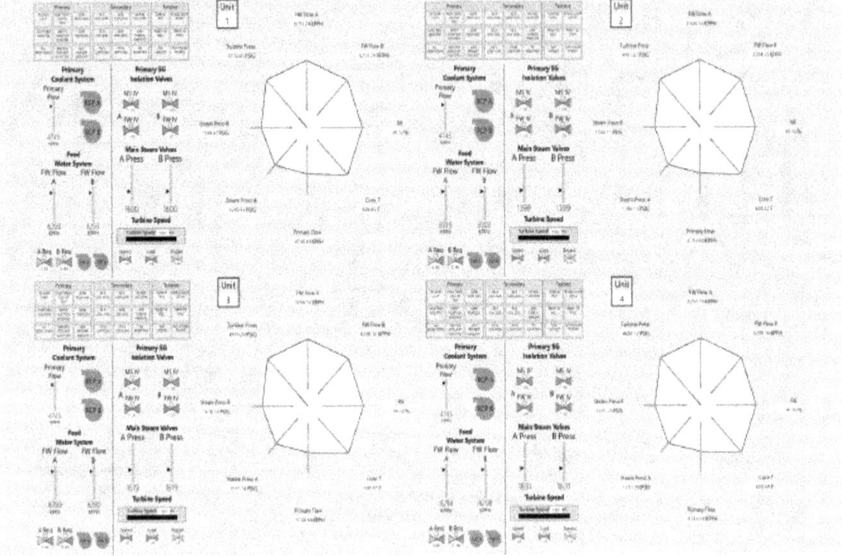

Figure 5.5: Updated Multi-unit Rancor Alarm and Summary Panel

"When" - Scenarios and Instructions

In this study, participants were tasked with working as a team to operate and monitor either a single Rancor plant with four units or two plants, each containing four units. The primary objective for these operators was to ensure the safety of the reactor plants, while also working towards load following as a secondary goal. In order to successfully complete the scenarios presented to them, the participants were required to start up and monitor the units, as well as manage any faults that occurred during the process. Each participant had specific roles and responsibilities, and the team had full autonomy to ensure plant safety. The scenarios presented to the teams represented the operational "when" of the ConOps - indicating the specific moments when the operators needed to interact with the Rancor system.

For Category 1 groups, participants were asked to work through four scenarios:

1. Startup procedure- ROs brought unit 3 online. The other units were programmed to be in a startup state upon commencing the study.

2. Steady-state operations- ROs varied controls to ensure the plant remained within appropriate ranges.

3. Planned Maintenance Shut Down Procedure - The SRO alerts the ROs that unit 1 needs to be shut down for maintenance activities.

4. Emergency Operating Procedure – Unit 2 experiences a fault. Loss of feedwater, feedwater pump trip.

For Category 2 groups, participants were asked to work through 4 scenarios:

1. Startup procedure- RO 1 brought unit 2 of plant 1 online and RO 2 brought unit 1 online of plant 2. The other units were programmed to be in a startup state upon commencing the study.

2. Steady-state operations- ROs varied controls to load follow and match demand, while ensuring the plants remained within appropriate ranges.

3. Planned Maintenance Shut Down Procedure - The SRO alerts RO 1 that unit 3 needs to be shut down for maintenance activities.

4. Emergency Operating Procedure –Plant 2, unit 4 experiences a fault.

In this study, the procedures for operating Rancor were provided to participants in a paper-based format. These procedures were developed by INL for each Rancor operational scenario, including start-up, shut-down, loss of feed water, and rapid shutdown. The paper-based procedures were an essential reference tool for the operators, providing them with step-by-step instructions for each scenario to ensure the safe and efficient operation of the power plant.

"Where" - Control Room Environment

The control room for this ConOps and study design consisted of three operators working in a single room. The control room featured three workstations, each equipped with dual monitors. The two Reactor Operators (ROs) had access to all human-machine interfaces (HMIs) and procedures relevant to their roles. Meanwhile, the Senior Reactor Operator (SRO) had access to all procedures and HMIs necessary for providing a general overview of either a single plant or both plants.

"How" - Description of Workflow

At the start of each trial in the study, participants were directed to their assigned workstations in the control room. The Rancor simulator was pre-installed on each workstation. For the first category of trials, the SRO prompted the participants to determine how they would divide control of the four units, and recorded their decisions for later analysis. Once participants were ready, they were instructed to begin scenario one and progress through all four scenarios. Throughout the trials, the researcher took notes on participants' think-aloud narratives, which were used to gain insights into their decision-making processes.

After completing all four scenarios, the study lead announced the conclusion of the experiment and participants were asked to complete a survey. While real-life control room operations typically involve pre-defined roles and protocols, the experimental design aimed to observe participant decisions and plant performance without strict adherence to a fixed order of operations, reflecting the exploratory nature of the research and the advanced concept of operations being studied. The study findings offer valuable insights into operator behaviors in an unbiased setting and provide a baseline for future research. With that said, to simulate real-world operations, participants were given two important behavioral constraints: they were required to communicate with each other throughout the study, and no action could be taken without agreement from all parties.

5.3   Study Protocol

The study protocol was designed to ensure that participants had a thorough understanding of the study procedures and could provide informed consent. Before the study began, participants were provided with a general overview training session on Rancor and the study. This training session lasted approximately 1 hour and included information on the experimental procedures and tasks. Participants were also given the opportunity to ask questions and seek clarification.

Each participant completed one trial of the study, which lasted about 90 minutes. During the study, participants were encouraged to communicate with one another and think aloud. This allowed the researcher to capture participant decision-making processes and assess mental demand. Participants were also allowed to ask questions if needed to ensure they had a complete understanding of the tasks.

After completing the study, participants were asked to complete a survey and participate in a post-experiment discussion, which lasted 30 minutes. The purpose of the survey was to gather feedback on the study procedures and tasks, while the post-experiment discussion provided an opportunity for participants to share their experiences and provide further insights into their stress and mental demand. All study procedures were approved by the Institutional Review Board (IRB), and informed consent was obtained from all participants prior to the start of the study.

Survey

Upon completion of the study, each participant was asked to complete an online Qualtrics survey consisting of qualitative and quantitative questions. These questions ascertained how the participants felt during the study and whether they felt they needed more, less, or the same number of team members were they to do this study again. The questions are given below:

Qualitative Questions:

- Could you have taken on more responsibility as an operator? Please explain.

- Did your stress increase at any point during this study? If yes, what caused the increase? Please explain.

- Were there enough operators to complete the tasks you were asked to do, yes or no? You have the option to provide details below your section.

- Would you have preferred if there were more operators? What about if there were fewer operators? Please Explain.

- Did you understand your job as an operator in this study, yes or no? You have the option to provide details below your section.

- What did you find frustrating during this study, if anything?

Quantitative Questions: Ranking Scale (0 to 5 Likert scale)

- How would you rate your mental demand throughout the study?

- How frustrated or stressed were you while completing a task?

- How calm and secure did you feel during the tasks?

- Do you agree with this statement: There were enough operators to complete the tasks you were asked to do.

- How effective were you at communicating as a team?

- Would additional crew members have been helpful in decreasing your stress?

- Do you agree or disagree with this statement: Additional Crew Members would have helped me complete my tasks as an operator.

5.4   Study Findings

This study aimed to address two critical questions: first, whether three operators could manage two plants with multi-units from a single control room without experiencing undue stress compared to three operators controlling a single multi-unit plant from one control room; and secondly, whether the participants perceived that there were enough operators to accomplish the required tasks. Encouragingly, the study's results indicate a positive response to both questions, supported by both quantitative and qualitative analyses. In the following sections, we will delve deeper into the outcomes and their implications for the remote control room's feasibility and efficacy in multi-unit plant management.

Quantitative Results

The findings of the quantitative survey are visually represented in Figure 5.6. This graph displays the responses to questions regarding the mental demand and stress levels experienced by the participants. Interestingly, both groups reported higher levels of mental demand and stress during the study. However, the difference in average scores between the two groups, namely the 2-plant and 1-plant groups, was minimal, with a range of only 0.5.

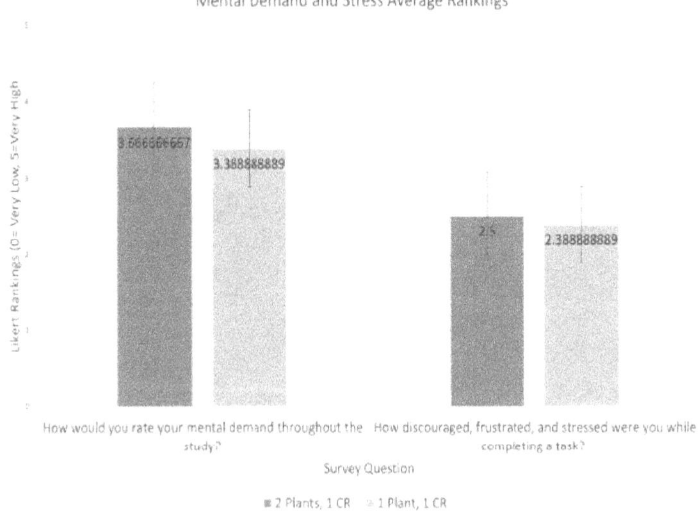

Figure 5.6: Survey results to Quantitative Questions 1 and 2

Interestingly, it is notable that all participants, regardless of whether they were in the 2 plant or 1 plant category, strongly agreed that there were enough operators to complete the tasks asked of them. When asked if additional crew members would help lower stress or complete tasks, participants in the 2 plant category showed a stronger inclination towards agreeing with those statements, though the difference between the two categories was less than 1. In the 2 plant scenario, there was one RO per plant and one supervisor shared between the two plants. In contrast, the 1 plant scenario had two ROs and one SRO. In the latter case, the ROs worked together, while the SRO provided general guidance as needed. As relatively inexperienced operators unfamiliar with Rancor, it is reasonable to assume that having a second operator in the single plant case helped to lower stress and complete tasks, which is evident in the lower averages of the one plant's case. Although all participants felt that there were enough operators, it is clear that having an additional operator would help lower their cognitive demand (see Figure 5.7). However, it is important to note that the conclusion should not be that more operators are necessarily required. The qualitative feedback revealed certain issues with the HMI that, if addressed, could help reduce the mental demand and stress of the operators.

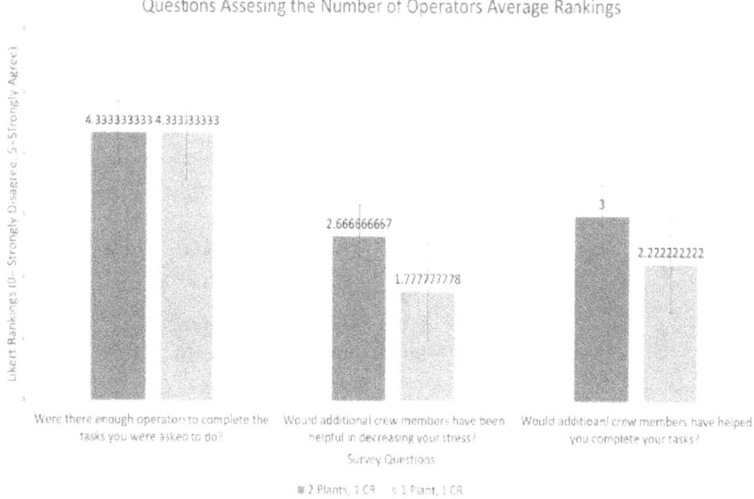

Figure 5.7: Survey Results to Quantitative Questions 4, 6, and 7

Qualitative Results

In the study, participants were tasked with starting up a unit, performing normal or rapid shutdown procedures, and carrying out steady-state operations, including load following. The Rancor simulator displayed the power produced by each unit and grid demand, which fluctuated periodically to simulate real-world scenarios. However, the timescale of the simulator was faster than reality, and participants had to adjust to changing demand from all four units throughout the trial. Qualitative feedback indicated that participants' stress and mental demand could have been reduced if the human-machine interface (HMI) and operator support tools were updated. Many suggested that a different screen layout, color changes, and more prominent alarm notifications would have helped.

Did your stress increase at any point during this study? If yes, what caused the increase? Please explain.

Participants reported increased stress levels when dealing with load following and feedwater faults, regardless of whether they were in the two plant or one plant category. This was based on the qualitative results obtained from the study. According to the written feedback, participants experienced stress in these situations due to a combination of factors, including their lack of familiarity with the system, the timescale of the simulation, and issues with the alarm presentation.

In particular, the limited familiarity that participants had with Rancor played a significant role in their stress levels. Some participants noted that they felt stressed when they first became familiar with the interface. This is not surprising, given that operator training is a critical part of licensing a nuclear reactor, and it takes time to learn how to operate a new system effectively. If participants had been given more time, even a week, to become familiar with the simulator and the controls, it is likely that their stress levels would have been alleviated.

Another factor that contributed to stress was the timescale of the simulator. The programmed demand changed rapidly, making it challenging for the operators to stabilize all four units. The rate of change when adjusting valves or reactivity was much faster than it would be in real life, which proved to be a stressor for the participants.

Specific feedback mentioned:

> "It felt stressful at times when trying to manage load/demand for the four units."

> "My stress level increased a bit when we had an alarm and we needed to follow the shutdown procedure. In our case, because we took some time to follow the instructions, when I looked at the screen the temperature was significantly increasing and we didn't have enough time to fix this, which lead to a reactor trip."

> "Yes. The increase was caused by having to closely monitor changing reactor conditions while also following instructions on how to respond to them.

In addition, the comments from participants highlight another issue with the timescale of the simulations and the procedures. The procedures used in this study were paper-based and followed a typical style used in nuclear power plant operations. These procedures required participants to first verify certain conditions before performing the necessary action steps. However, this structure does take time to read through and enact. During the study trials, every participant encountered a situation where they were reading through a procedure while the simulator parameters were adjusting, leading to a trip occurring before the participants were able to take a control action. This issue can be attributed to a combination of factors, including unfamiliarity with the system, the style of the procedures, and the fast-paced timescale of the simulator.

Overall, the feedback suggested that stress levels could be reduced by improving operator training and providing more time for familiarization with the simulator. Additionally, adjusting the timescale of the simulator and improving the alarm presentation could also help alleviate stress levels during the simulation.

Were there enough operators to complete the tasks you were asked to do, yes or no? You have the option to provide details below your section.

All participants responded positively to this question, which aligns with the findings from a similar question in the Quantitative section. The feedback provided can be summarized by the following quotes:

> "I think one operator could reasonably manage these responsibilities with some supervisory assistance."

> "I think managing the 4 reactors was not too difficult, especially if I had proper training it would be a lot easier to get into a flow."

> "I think having one other operator to verify what I am doing and the alarms, was helpful but the third person wasn't really necessary."

> "1.5 operators for 4 units was good."

Participants stated that with increased familiarity and a supervisor to assist in monitoring and managing alarm conditions, it would be feasible to control either one plant or two plants from a single control room.

Would you have preferred if there were more operators? What about if there were fewer operators? Please Explain.

The responses to this question varied slightly. Participants in the one plant case seemed to be satisfied with having two operators and one SRO. However, participants in the two plant category felt that one operator and a shared SRO were sufficient but would have preferred if the screens displayed information differently. The following quotes summarize the feedback:

> "I think 1.5 operators for 4 units was more than enough. 4 units can really be handled by 1 operator. However, the 0.5 operator came in handy when there was an emergency (feed water failure)."

> "1.5 operators were plenty. Due to my inexperience operating reactors, I felt busy and even somewhat frantic at moments. However, near the end I got the hang of it and felt like 1 was fine."

> "Three operators are perfect, but no more than that. Potentially one operator is more than enough."

> "I think two operators would be the ideal, one for the controls and one for the monitoring. As long as both operators are able to function properly according to their respective responsibilities, I believe that two is a good number. Having three operators would be okay but not necessary."

> "I think two operators for is adequate for four units. Having a second person to monitor felt it was crucial given so many units to keep track of."

> "Too much information on the screen for just one person. Two operators would be better."

> "I think having a solo operator and having someone looking over your shoulder was appropriate for monitoring. However, having two screens made it quite difficult to monitor errors."

In the one plant case, one RO performed control actuations while the second RO monitored the alarm panels and critical values such as core temperature. In the two plant case, one RO had to perform control actuations and monitor the alarms. In both cases, the SRO acted as a helping hand ready to guide the operators when needed and helped monitor the alarms. Based on the survey results, it is clear that the second RO and the SRO were mostly needed for alarm monitoring. The feedback indicates that the alarms were quite challenging to monitor due to their design. The font was small, and the layout was difficult to follow. Since all participants felt there were enough operators at any given time, and some noted that even fewer operators could be manageable, it is likely that if the Rancor HMIs were adjusted to reduce mental demand and stress, there would be a decrease in the desire for additional operators. In a future study, it would be interesting to see how the preference for additional operators may change if the alarms and layout of Rancor are adjusted to reduce the mental demand on operators and they feel they can complete tasks mostly on their own.

What did you find frustrating during this study, if anything?

The operators expressed some consistent frustrations with the HMI and simulator, which mainly centered on the design of Rancor itself:

> "It is difficult to find the right buttons to press because the controls are separate from the visuals of the components."

> "The layout is difficult to navigate."

> "The font is too small in several places on the HMI."

> "The alarms are difficult to read. The font is too small and the way the alarms are organized made it hard to locate problems."

> "There was frustration with how quickly the simulator responded to slight adjustments. At times, the rate of change is too quick for operators to respond."

Overall, the participants' frustrations were related to Rancor's design, rather than the number of operators or plants being controlled. This suggests that if the operators were given updated HMIs and the simulator's rate of change were adjusted, they might not feel that additional operators would reduce their stress or mental demand.

5.5 Rancor Simulator Data

Data visualization plots were produced from the simulator data to better correlate mental demand to what was happening within the simulator during a study trial. Figure 5.8 shows the simulator data for plant 1 in the two plant scenario, showing the fluctuations in reactivity, core temperature, load, and bypass valve. The participant had to bring Unit 2 online, adjust reactivity and core temperature during startup, and then match demand on all units. The fluctuations in the load and bypass valve plots represented changes made by the operator to adjust the reactor power output based on grid demand. To accommodate the decreased load output, the participant lowered the reactivity of all four units, and manipulated the bypass valve to maintain core temperature within the appropriate operational bounds. After about 20 minutes, this participant was asked to shutdown unit 3, leading to a drop in the reactivity, load output, and core temperature for this unit. The bypass valve was increased to aid with cooling the reactor unit. The whole study trial required diligent and continuous monitoring of changing plant parameters and frequent adjustments, which likely contributed to the participants' high levels of stress and mental demand.

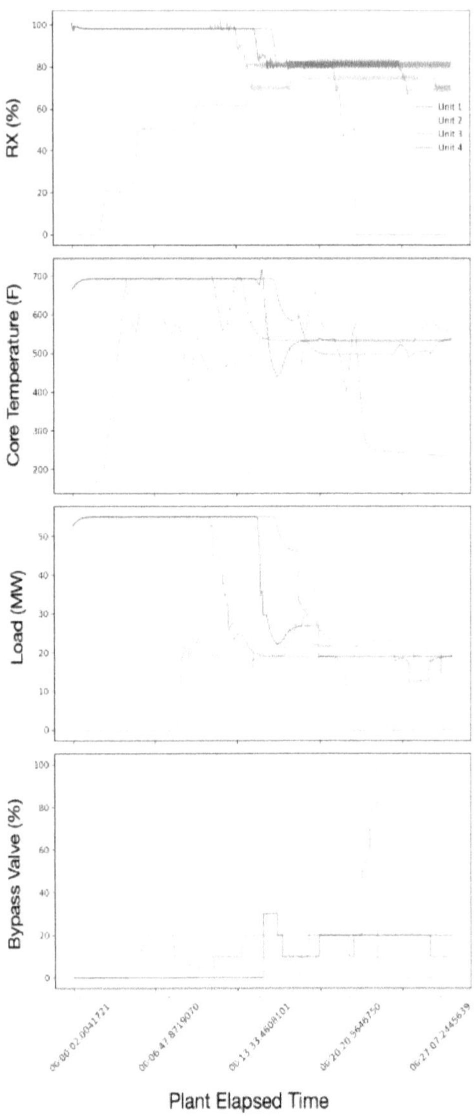

Figure 5.8: RO1 Plant 1, 2 Plant Startup Scenario Simulation Plant Data

Mental demand was likely heightened when a unit tripped due to exceeding the operational threshold limit, adding to the already challenging task of operating the plant. Figure 5.9 provides an illustration of the plant parameters during a trip event. During this incident, units 1, 3, and 4 experienced a reactor and turbine trip. As shown in the core temperature plot, all three units reached a temperature of over 700°F, causing the turbine to trip and resulting in the load plots showing 0 MW. To lower the core temperature, the bypass valve was opened to 100% in all three cases. Despite the SRO's assistance in mitigating the trips, participants reported feeling significant stress due to the unfamiliarity with the system, which was compounded by the need to monitor and control all other units while operating within the simulator's compressed time frame. In some cases, attempting to mitigate a trip on one unit caused other units to trip, further increasing stress levels. Overall, the qualitative feedback and data visualization suggest that updating the HMI and operator support tools could alleviate some of the cognitive load on operators. Importantly, the study did not conclusively demonstrate that more operators are required.

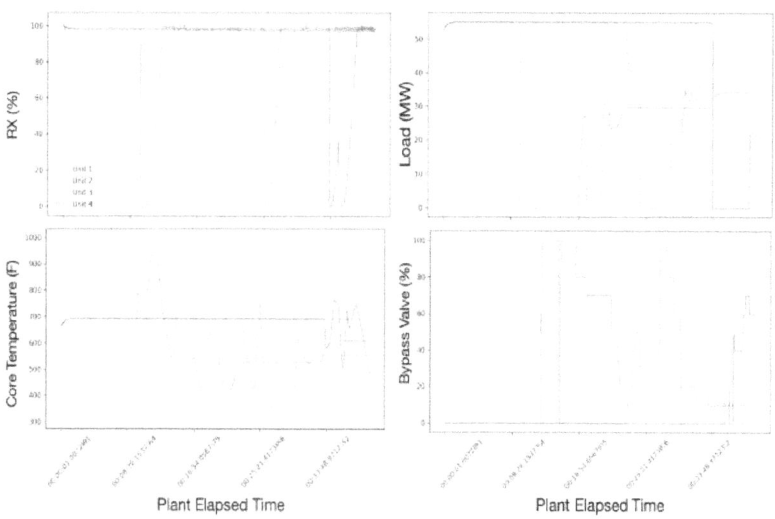

Figure 5.9: Single Plant Scenario Simulation Data

5.6 Summary and Key Findings

Although the study does not provide conclusive evidence, it does indicate that two multi-unit nuclear power plants can potentially be operated from a single control room without an increase in the number of operators. The research findings suggest that three operators can efficiently control two plants with multiple units from a single control room, without experiencing significantly higher stress levels, compared to controlling a single multi-unit plant. Nonetheless, the study participants encountered stressors related to the HMI design and the backend of the Rancor simulator. Despite limited familiarity with the system, the operators were able to complete their tasks, and all participants agreed that there were sufficient operators to manage the scenarios. These results imply that mental demand and stress could be further reduced through updates to the system's design and increased levels of automation. For instance, automating the manual load-following procedures could help alleviate operator stress.

Building upon these findings, advanced reactor companies can consider exploring the concept of remote operations using a reduced number of operators to control multiple plants within a single control room. This can be achieved by leveraging advanced digitization to enhance operator communication, collaboration, and control efficiency.

The specific ConOps used in this study, which utilized three operators to control two plants from a single control room, provides a starting point for further research. By refining the use of advanced HMIs and optimizing communication between operators, companies can iterate on this design and explore the potential of using fewer operators to control even more plants within a single control room. In addition, the integration of computer-based procedures (CBPs) can further improve control efficiency and safety. The use of CBPs can automate the execution of specific procedures, reducing the need for manual input and potentially decreasing the likelihood of errors. The use of digitization and CBPs can also enable operators to easily access critical information related to plant history, safety limits, and operational procedures.

Overall, this study highlights the potential for remote operations to enhance the scalability and cost-effectiveness of advanced nuclear reactors. Companies should experiment with different combinations of operators and plants within a single control room, continually iterating the control design until an appropriate HMI design and level of automation is found that enables the control of several nuclear power plants from a single remote location.

Chapter 6

Conclusions, Insights from Experts, and Remote Operations Criteria

Based on the research conducted in this dissertation, it can be concluded that technological advancements and the emergence of new reactor designs have made a remote control operational strategy a possibility. The dissertation successfully demonstrated the technical feasibility of remote operations in the nuclear industry by controlling the Kairos Power ETU system from a remote control room. Furthermore, the human factors study outlined in chapter 5 also baselined the possibility of controlling multiple plants from a single control room, which would be a viable option if remote control rooms were adopted by advanced reactor companies. These findings suggest that remote operations could provide a more efficient approach to nuclear power plant operations, with the potential to improve both safety and cost-effectiveness.

Remote operations can offer significant benefits to the nuclear industry, including reductions in operations and maintenance costs and enhanced scalability of nuclear power. By enabling geographically dispersed plants to be operated from a centralized location, remote operations can reduce the need for on-site staffing and associated costs. This could lead to substantial cost savings, making nuclear power more competitive in an increasingly cost-sensitive energy market.

Furthermore, remote control rooms increase the scalability of advanced nuclear power plants. By allowing for multiple plants to be operated from a single location, remote operations enable more efficient and effective management of resources, which could lead to cost savings and better utilization of equipment and personnel. This could make nuclear power more attractive to investors and provide a pathway for future growth of the industry.

6.1 Insights from Industry Experts: The Value and Challenges of Remote Operations in the Nuclear Industry

In order to gain insight into the practical considerations and potential challenges surrounding the implementation of remote operations in the nuclear industry, this dissertation included interviews with three high level employees from Kairos Power. The interviewees included an instrumentation and controls expert, a licensing specialist, and a strategist, each with valuable perspectives on the subject.

In these interviews, the Kairos Power employees shared their views on the value of remote operations, highlighting potential benefits such as increased safety, improved reliability, and cost savings. They also discussed the potential challenges and obstacles to implementing remote operations, such as regulatory barriers, cybersecurity concerns, and cultural resistance to change.

It is important to note that the responses from the Kairos Power employees have been summarized and presented in an aggregated format in this dissertation, without the use of direct quotes. This is because the intention of the interviews was to gather insights and perspectives, rather than to present a specific stance or position from Kairos Power as a company on remote operations. By presenting the views of these high-level employees in a summarized format, this section aims to provide a general understanding of the potential benefits and challenges associated with remote operations in the nuclear industry, as well as key considerations for its implementation. These insights can be used to inform future decision-making and regulatory development related to remote operations.

Overall, this section serves as a valuable addition to the dissertation by providing practical insights into the perspectives of experts in the industry, while also emphasizing the importance of considering a range of perspectives and challenges in the implementation of remote operations in the nuclear industry.

Instrumentation and Contorls Expert

Instrumentation and controls (I&C) is a specialized branch of engineering that primarily deals with the measurement and control of process variables. It involves the design and implementation of highly sophisticated systems that incorporate these variables to optimize performance and ensure efficiency. The director of I&C at Kairos Power was asked about the value of remote operations for the nuclear industry, and their response centered on the issue of scalability. Although nuclear power is an incredibly potent energy supplier with high energy density and output, it faces challenges when it comes to scalability and adaptability to changing energy grids. For instance, large LWRs that produce 1,000MWe are not easily capable of load following or adapting to the introduction of renewable energy sources. The solution to this lies in the design of SMRs, which are better equipped to handle load following situations and can be built more quickly than their predecessors. Furthermore, to achieve greater operational efficiency and effectiveness, it is necessary to coordinate the operation of multiple units from a centralized location, a standard practice in the power generation industry. Remote operation is a key aspect that needs to be figured out to ensure successful scalability.

The main obstacles to implementing remote operations, according to the director, are regulatory issues and public perception. Nuclear power has a history of safety concerns, which have led to a general bias against it. This bias makes it inherently challenging to implement remote operations. The public perception of remote operations is that they are unreliable and subject to extreme cyber threats. This perception is even more heightened when it comes to nuclear energy. However, from an I&C engineering perspective, the solution to the perception obstacle lies in separating the protection systems from the control systems. By proving that no control malfunction, whether due to technicalities or malicious actors, can affect the protection system, the perception problem can be resolved. Therefore, rather than focusing solely on scrutinizing the remote control room, more attention should be given to scrutinizing the reactor protection system.

Licensing and Regulatory Framework Expert

Those who work in licensing for nuclear reactors are required to have an in-depth understanding of the NRC regulations. From a regulatory licensing perspective, the primary obstacle to remote operations is the regulatory body itself. The expert interviewed expressed concern that not only is there no regulatory framework for this operational strategy yet, but there is also uncertainty regarding whether the NRC would support it. It is evident from the interview that any advanced reactor company that pursues a remote operations control room should expect significant pushback from the regulatory body. However, this pushback can be overcome.

Remote operations is not new, but it is for the nuclear industry. Therefore, part of the strategy for advanced reactor companies must involve getting the regulatory body more comfortable with remote operations for advanced nuclear power plants. To achieve this, a company would need to demonstrate the strength of the reactor protection system and ensure that it is decoupled from the control strategy. Additionally, one needs to consider operator licensing for a remote control room and continuous engagement with the NRC. Nuclear power plant operator licensing has always been for on-site control rooms. This paradigm shift will require exploring new ways of training operators, and engagement with the NRC will be vital for socializing the concept of remote operations and not surprising them with a new operational strategy late in the licensing process.

Strategy & Innovation Expert

Developing a successful strategy for advanced nuclear reactors involves not only designing and building the reactors, but also effectively selling them to vendors, the public, and the Nuclear Regulatory Commission. The Vice President of Strategy and Innovation at Kairos Power knows that one key area where strategy can have a significant impact is in reducing operations and maintenance costs, and remote operations can play a crucial role in achieving this goal. Small modular reactors are particularly well-suited for remote operations due to their size and design simplicity, which allows for a more decentralized operational strategy. However, there are several significant obstacles to implementing remote operations in the nuclear industry, including policies, regulatory requirements, and public perception.

The history of the nuclear industry has shown that accidents often lead to increased regulations, making it more difficult to introduce new technologies and operational strategies. The safety culture surrounding nuclear power is strict, and convincing regulators and the public to embrace remote operations will require careful consideration and strategic planning. Hence this strategy expert stressed that while Gen IV reactors are designed to be intrinsically safe and highly automated, simply emphasizing the separation of reactor protection systems and control systems may not be enough to win public support. Instead, a successful strategy will need to take into account the importance of human involvement in the operation of these advanced reactors, and ensure that the automation is designed with the needs and preferences of operators in mind.

According to this interviewee, to successfully implement remote operations for advanced nuclear power plants and win over regulators and the public, the industry will need to continually showcase the safety and effectiveness of this strategy. This will involve ongoing refinement of the remote control room design, communication infrastructure, and automation systems, as well as a concerted effort to educate and engage the public on the benefits of advanced nuclear reactors and their operational strategies. Ultimately, a successful strategy for remote operations will require collaboration and communication between industry leaders, regulators, and the public to build trust and achieve a shared understanding of the benefits and challenges of these new technologies.

6.2   Initial Criteria for Successful Implementation of Remote Operations in Advanced Reactors

Based on the summary of the interviews, it is clear that there are several key criteria that advanced reactor companies must address to successfully implement remote operations. Firstly, it is important to have a clear separation between the reactor's protection systems and its control systems. This separation ensures that there is minimal risk of interference between the two systems. This is critical because safety-related control functions are designed to prevent accidents and protect the public from the harmful effects of radiation in case of any operational anomalies or accidents. Thus, these safety functions must remain local to the plant and have complete autonomy, ensuring that any potential safety risks are addressed immediately without relying on external systems or remote operators. On the other hand, non-safety related control functions, such as routine plant operations, do not directly impact plant safety. Therefore, they can be remotely operated or controlled from external locations. By ensuring a clear separation between these functions, the safety and reliability of the nuclear power plant can be maintained, minimizing the risk of accidents and protecting public safety.

In addition to this, it is important to have an optimized level of automation that works well with human operators. This means that the automation should be designed in a way that complements human decision-making, rather than replacing it entirely. This helps to ensure that operators are able to effectively monitor and control the reactor, and respond to any incidents or emergencies that may arise. To achieve this optimized level of automation, it is important to conduct human factors studies and iterative evaluations to determine the appropriate level of automation for the given task and environment. This involves a continuous feedback loop where the system is evaluated, tested, and refined based on the input of human operators. Furthermore, it is important to consider the role of computer-based procedures (CBPs) in facilitating remote operations. CBPs can help to streamline and standardize procedures, reducing the potential for human error and enabling more efficient operations. However, it is important to ensure that the CBPs are designed in a way that complements human decision-making and does not lead to overreliance on automation. By considering the human factors implications of remote operations and conducting iterative evaluations, advanced reactor companies can optimize the level of automation in a way that enhances operator performance, safety, and efficiency. This will help to ensure the successful implementation of remote operations in the nuclear industry and maximize the potential benefits of this emerging technology.

Another key consideration is the development of new training programs for operators of remote control rooms. These programs should be designed to help operators gain the necessary skills and knowledge to effectively operate and monitor the reactor from a remote location. Early engagement with regulators is also important. Companies should begin talks and meetings with the regulators as early as possible, and involve them in the design process. This helps to ensure that the regulator's concerns and requirements are taken into account from the very beginning, and can help to facilitate a smoother regulatory approval process.

Finally, it is important to continually test and demonstrate the technology, and to constantly look for ways to improve it. This can involve conducting multiple small-scale tests and demonstrations, as well as seeking feedback from operators and other stakeholders to identify areas for improvement. By taking a continuous improvement approach, companies can help to build confidence in the remote control technology, and ensure that it is safe, effective, and reliable.

List of Criteria for Remote Operations Consideration:

In addition to the NRC ground rules, and encompassing the feedback from the interviews, there are several criteria that must be considered when implementing remote operations in the nuclear industry. These criteria are summarized here and will need to be carefully evaluated and addressed in order to ensure safe and effective remote operations:

Communication and Cyber Security:

- Communication infrastructure between the control room and the reactor, as well as between the remote control room and any local control rooms, must be in place and highly reliable. The communication should also be secure to prevent any unauthorized access or interference.

- Strict and redundant cyber security protocols must be implemented to protect all safety-related functions of the reactor.

Safety-Related Functions:

- Complete separation of safety-related functions from the control system is a fundamental requirement to prevent potential safety issues caused by control system malfunctions.

- Safety-related control functions, whether automated or not, shall remain local to the plant and shall have complete autonomy over non-safety related control functions.

Access Control and Protocols:

- Additional protocols shall be implemented to ensure that the location of remote control commands is validated, meaning that the control system can verify the physical location from which the commands are being sent. This helps prevent unauthorized access to the control system and ensures that the commands are being sent from a trusted source.

- Additional protocols shall be implemented to ensure that the authority of the command source is authenticated, meaning that the control system can verify that the commands are being issued by someone with the appropriate level of authorization. This helps prevent unauthorized access to the control system and ensures that only authorized personnel can issue control commands.

- Interactive permissioning should be in place to ensure that only authorized personnel have access to the digital control room information, with different levels of access depending on their role.

- Protocols must be in place in case the remote control room is no longer able to control the reactor or reactors.

Regulatory Compliance:

- Continuous discussions with the regulatory body are necessary to ensure compliance and address any concerns.

Iterative Design and Evaluation:

- Iterative design of the system is essential, based on evaluations and human factors studies to determine the appropriate level of automation, the most effective HMIs and OSSs, and the optimal operational strategy that aligns with human operators.

- A comprehensive Concept of Operations (ConOps) should be developed that encompasses all of the criteria listed and considers exactly how operators will interact with the reactor(s) they control. The ConOps should be iteratively evaluated alongside individual components such as the HMI and OSSs.

- Clear roles and responsibilities for the operators, a clear communication protocol between operators, a clear plan for how operators will interact with the reactors they control, and a clearly outlined path in case of an emergency scenario are all critical components of a successful ConOps. Evaluate the HMIs, OSSs, and operator roles, among other things, in the context of multi-unit and multi-plant operations, if they will be included in the ConOps. This testing and evaluation will help ensure that the remote operations system is able to effectively manage multiple reactors and plants from a single remote location while maintaining a high level of safety and reliability.

Operator Training and Responsibilities:

- A new operator training program for remote operators is required based on functions operators will be able to control remotely.

- Clear roles and responsibilities for the operators, a clear communication protocol between operators, a clear plan for how operators will interact with the reactors they control, and a clearly outlined path in case of an emergency scenario are all critical components of a successful ConOps.

Advanced reactor design organizations can progress towards remote operations by thoroughly assessing and meeting these criteria, which serve as a starting point. Similar to the NRC ground rules, this list is dynamic and will likely expand as control rooms and operational strategies for advanced reactors evolve through iterative design processes.

6.3 Concluding Remarks and Future Work

The research presented in this dissertation demonstrates the technical feasibility of remote operations in the nuclear industry, presents a potential use case of a remote control room, and highlights the benefits and the challenges of using digital remote control rooms for advanced nuclear power plant operations. By providing a comprehensive set of initial key criteria, in addition to the NRC's ground rules, that must be met for successful implementation of remote operations, this research offers valuable insights to industry professionals as they navigate the future of nuclear power.

Furthermore, by conducting a human factors study that compared the mental demand of operating two plants from a single control room versus one plant from a single control room, this dissertation also provides insights into the design considerations for remote control rooms for advanced nuclear reactors. The study highlights the importance of optimizing operator performance and minimizing cognitive load to reduce operations costs. This involves considering factors such as display layout, control room layout, alarm management, and operator responsibilities.

The findings of this dissertation may also offer insights into the public acceptance of remote operations for advanced reactors, which is crucial in light of historical public concerns about nuclear power safety and environmental impact. The implementation of remote operations has the potential to enhance safety, simplify operations, and increase scalability of nuclear power. By demonstrating these potential benefits, continuously testing and demonstrating remote operations strategies, and engaging with stakeholders to address their concerns in a transparent and open manner, the industry can work to build trust and foster public acceptance of this technology.

Moving forward, there is a clear need for rigorous and iterative testing and design of the digital remote control room, safety systems, cyber security protocols, and automation levels. The demonstration of remotely operating the ETU as detailed in Chapter 4 and the use case study baselining the operations of two plants from a single control room described in Chapter 5 are two initial steps in the open iterative design process. Collaboration with cyber security experts will be critical in developing and integrating advanced protocols to protect the system from potential threats. Ongoing testing and demonstration of the technology will also be necessary to build confidence in its safety and reliability. However, it is important to note that public acceptance of remote operations is not guaranteed. Therefore, the industry must accompany the implementation of this technology with a strong communication and engagement strategy to ensure that the public understands the benefits and risks associated with this approach.

In conclusion, the findings of this dissertation provide valuable insights into how the industry can successfully implement remote operations while addressing the concerns of regulators, operators, and the public. As the nuclear industry considers the future of power plant operations, the findings from this research should be taken into account. By embracing the potential benefits of remote operations, addressing the challenges head-on, and continuously testing and demonstrating the technology, the industry can work towards playing a vital role in meeting our future energy needs.